basics of
missile guidance
and
space techniques

$3.90

basics of missile guidance and space techniques

by MARVIN HOBBS

VOL. 1

PREFACE

In general, the development of missile guidance has encompassed a broad range of specializations ranging through rocket dynamics, hydraulics, gyroscopics, and electronics. The development of short-range guided missiles has led to the perfection of radar tracking and command techniques as well as various forms of terminal homing. The development of operational long-range guided missiles has demanded the perfection of ballistic guidance techniques of high precision, combining gyroscopics and electronics. Some of this capability has been displayed to the world by the launching of Earth satellites, even though in many instances they have not indicated guidance capability of a high order of accuracy.

The further development of satellite and space exploration operations opens a new field, which we have chosen to call *space techniques*. Without good guidance techniques it is impossible to place satellites even close to desired orbits. Without space electronics it would be impossible to track, control, and bring together the necessary elements for travel beyond an Earth-satellite orbit.

With the realization of flight beyond the direct range of the Earth's gravitational field there comes the need for a whole new realm of communications, data transmission, and navigational aids operating across the great distances between satellites, planets, and the stars. Thus, rocket guidance and related space techniques will go forward until man succeeds in the conquest of space. The fundamentals of missile guidance are already well established, and are being applied daily throughout the world.

The concepts of space electronics have already been applied in radio astronomy and satellite programs. They are being developed now for their eventual application in precise satellite orbits, trips to the moon and in interplanetary exploration. Even before the outer space operations are widely developed, men should find satellites useful in inner space as relays for microwave signals spanning the globe in world-wide television, and multichannel telephony circuits, and as international message carriers. In addition, the orbital bases will prove useful for navigational reference as well as reconnaissance, mapping, and weather forcasting stations.

The close relationship of space systems, already beginning to evolve for both military and commercial purposes, and missile development, highlights the need for their treatment in the same context. These volumes were written to provide a comprehensive basic course in missile guidance and related space systems in fundamental terms. They are directed to the educational needs created by the rapidly growing requirement for technical and semitechnical personnel to satisfy present and future missile and space vehicle programs, as well as the education and enlightenment of those desiring to learn more about the why and how of missile guidance and space techniques on an informal basis.

The author wishes to thank and credit the following firms, sources, and agencies: American Rocket Society; Institute of Radio Engineers; American Bosch — Arma Corp.; Kearfott Co.; Fairchild Controls Corp.; Citizenship Radio Co.; Stepper Motors Corp.; *Missile and Rocket Journal;* The Information Offices of the Departments of Defense, Army, Navy, and Air Force; The National Aeronautics and Space Agency; *Space Handbook, Next Ten Years in Space, 1959-1969,* and other U. S. Government publications of the House Space Committee; *History of German Guided Missile Development,* prepared for NATO; the British Institute of Navigation, the British Information Services, and Zenith Radio Corp.

The author also wishes to thank the managements and staffs of the General Instrument Corp. and of John F. Rider Publisher, Inc., for their consideration, and for the assistance they rendered in the many details essential to the preparation of the manuscript for publication.

In addition, to the valuable assistance given by the above, specific credit is due the following: page 2-95 — *TIME* Diagram by R. M. Chapin, Jr. Reproduced by permission of *TIME,* copyright TIME INC., 1959; pages 2-25 and 2-26 — Photographs from the Mount Wilson and Palomar Observatories; page 1-139 — After *Applied Hydraulics & Pneumatics,* September 1958; pages 2-83, 2-108, 2-118 and 2-119 — Photographs from *Astronautics,* May 1959.

Chicago, Illinois.   MARVIN HOBBS
June 1959

CONTENTS

VOL. 1 — BASICS OF MISSILE GUIDANCE
AND SPACE TECHNIQUES

Introduction to Missile Control and Guidance

In about two decades both short-range and long-range guided missiles have evolved from simple unguided rockets, autopiloted aircraft, and the basic radar systems for aircraft detection and tracking. Practically all types and ranges of missiles require guidance to be effective against either moving or fixed targets. (The term *fixed targets* means those fixed to the Earth. From a space viewpoint all targets are moving. For instance, while a ballistic missile is in flight the target at which it has been fired moves an appreciable distance because it is fixed to a rotating Earth.)

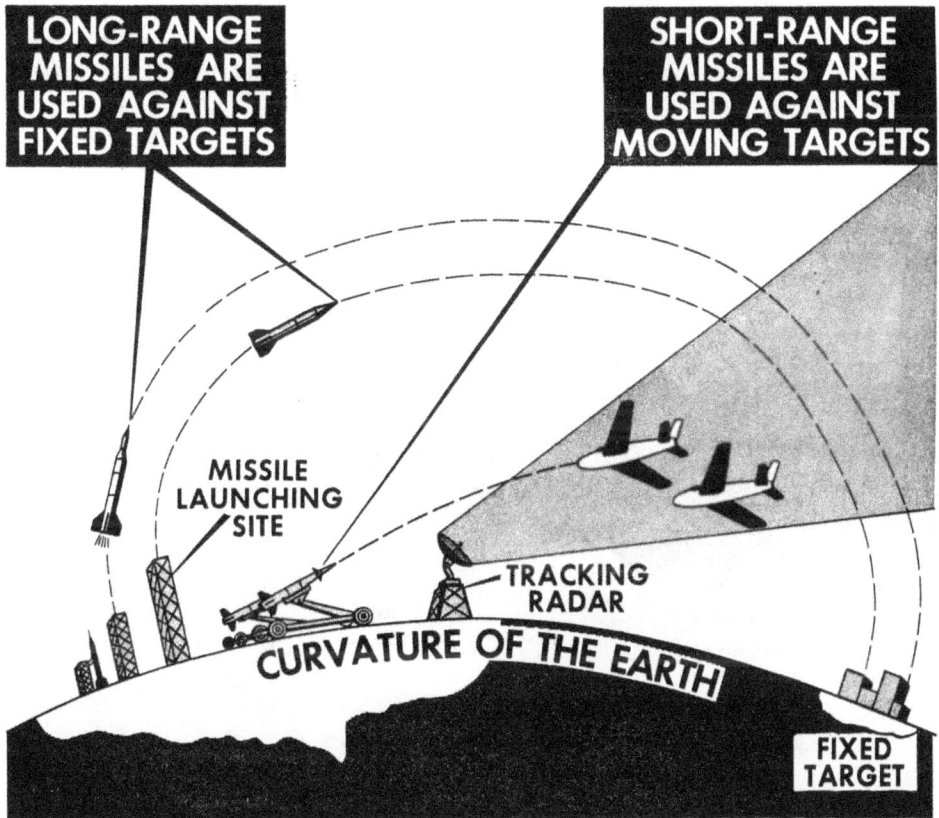

Long-range missiles are used against fixed targets over distances ranging from 200 to 5000 miles, and require a high order of navigational accuracy to insure that the error at the target area will not be excessive. Short-range missiles may be used against either fixed or moving targets in ranges less than 200 miles. Since most short-range missile applications are against moving targets, a high degree of target observation is necessary, and a rapid-action system for bringing the missile to bear on the target is essential.

THE EVOLUTION OF GUIDANCE TECHNIQUES

Remote Control by Wire

One of the earliest short-range guided missiles was the German X-4 of World War II. Control signals were transmitted over a pair of wires from the mother aircraft to this air-to-air missile. More than three miles of .2 mm insulated steel wire was reeled from the wing tips of the missile as it sped toward a hostile bomber. The position of the missile relative to the target was observed optically from the launching aircraft, and it was directed onto the desired trajectory without any form of radio or radar transmission. Significant information was sent over the wires to control the missile about two axes. These control signals were sent either as d-c variations or as a-f signals. In the d-c system, one set of orders was transmitted by switching the d-c polarity and the other set by switching two levels of current. In the a-f system frequencies above and below reference values were sent to provide the appropriate orders.

In another missile development, designated the X-7, the Germans considered using this type of guidance for attacks against tanks and other ground vehicles from ground launching sites. Since then in other countries this type of guidance for anti-tank missiles has received further attention and in the U.S.A. a wire-guided missile, known as the Dart, was under development at one time. Wire-guidance is characterized by its simplicity as well as its immunity to jamming and electronic countermeasures.

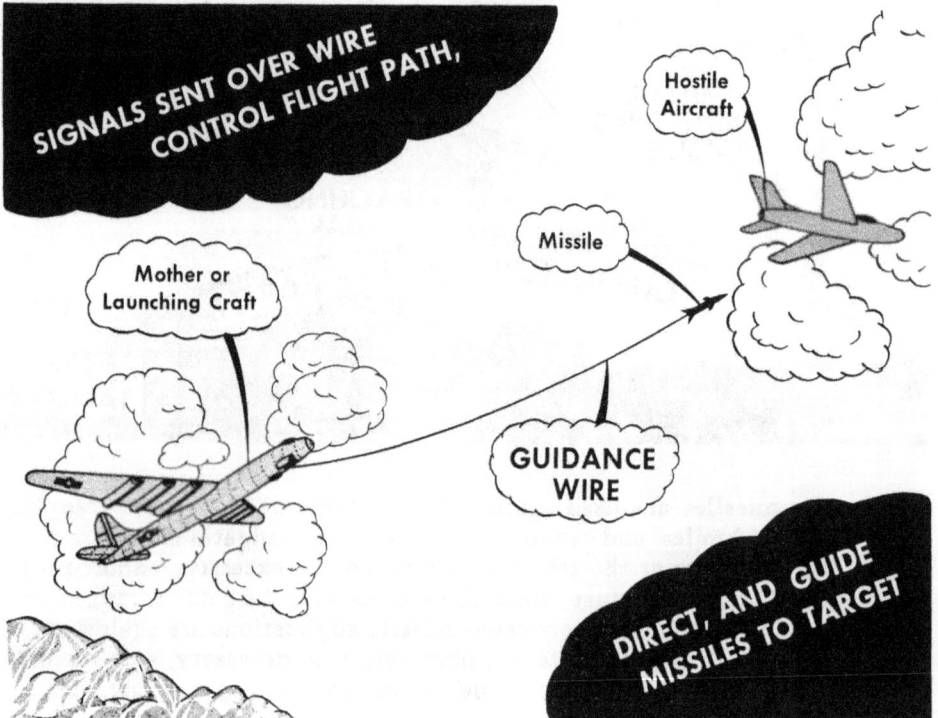

SIGNALS SENT OVER WIRE CONTROL FLIGHT PATH,

Hostile Aircraft

Missile

Mother or Launching Craft

GUIDANCE WIRE

DIRECT, AND GUIDE MISSILES TO TARGET

Remote Control by Radio

One of the simplest forms of remote control by radio signals is that practiced by model-aircraft enthusiasts. By transmitting amplitude-modulated pulses from a ground-radio transmitter to the receiver in the model, the flight-control elements are adjusted in a step-type operation. Smoother, and more continuous control can be exercised with potentiometers or miniature servos responding to the commands detected in the model plane. The aircraft is observed visually, and is maneuvered as desired by proper signals under the operator's control. More sophisticated versions of such radio techniques have been utilized to control drone-aircraft used as targets for missile testing. It is obvious that such a simple system would be easily susceptible to jamming, but it can be very useful where such considerations are not involved.

REMOTE CONTROL IS ACCOMPLISHED BY RADIO CONTROL SIGNALS

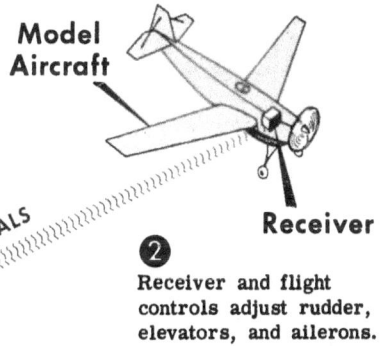

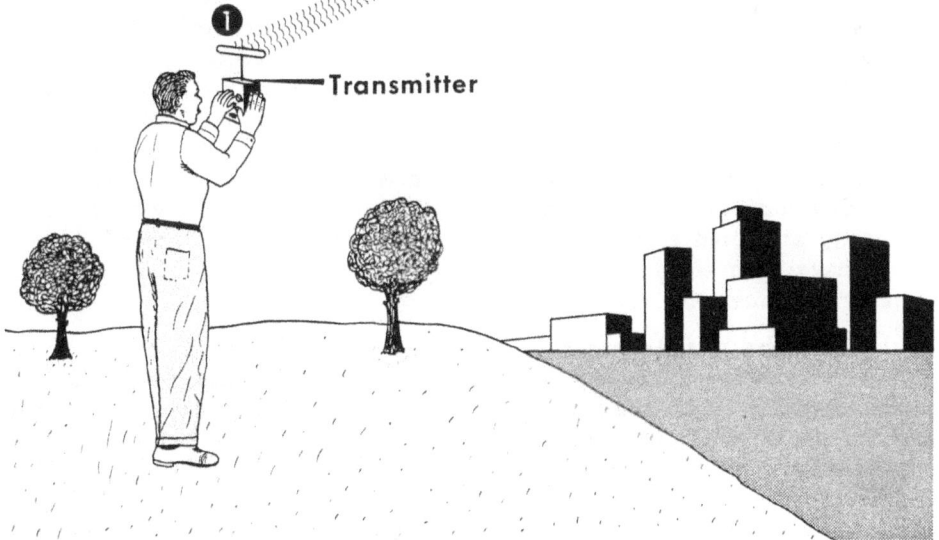

Model Aircraft

Receiver

A-m pulsed radio signals from ground transmitter cause the model aircraft to maneuver as desired.

RADIO CONTROL SIGNALS

❷ Receiver and flight controls adjust rudder, elevators, and ailerons.

❶

Transmitter

Remote Control by Radio (contd.)

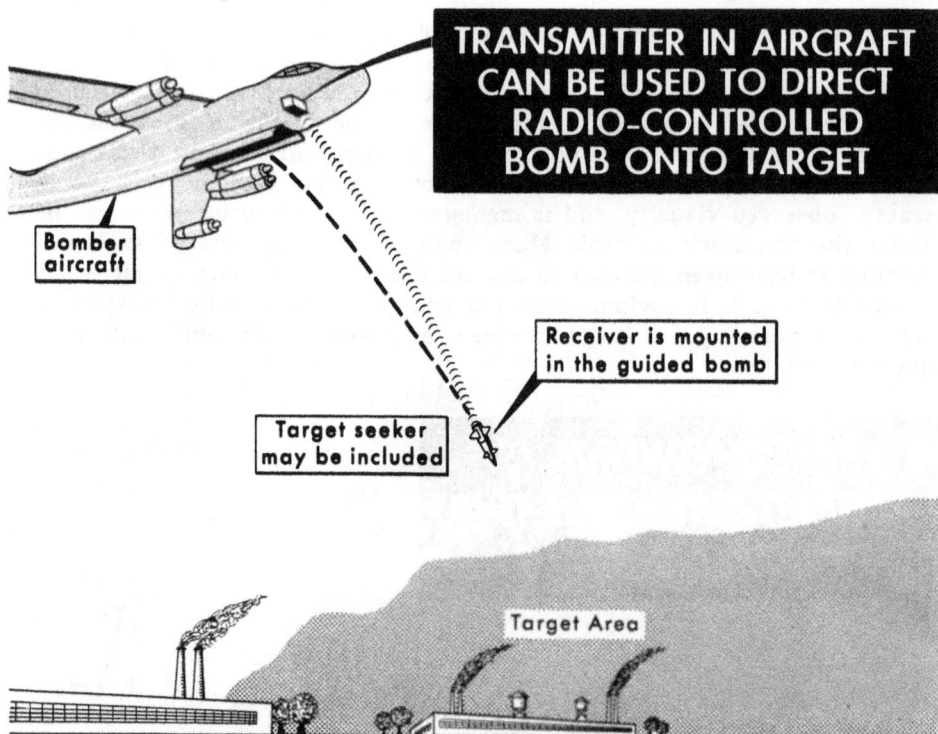

TRANSMITTER IN AIRCRAFT CAN BE USED TO DIRECT RADIO-CONTROLLED BOMB ONTO TARGET

Bomber aircraft

Receiver is mounted in the guided bomb

Target seeker may be included

Target Area

We can also place the control transmitter in a military aircraft to enable us to direct the bombs by means of radio signals to the target. During World War II the U.S.-developed Azon and Razon bombs were directed in this manner in azimuth, and both range and azimuth, respectively. The Azon bombs were used operationally for some of the more accurate bombing operations against bridges and road transport facilities both in Europe and the Far East. Also, the Germans placed into operational use two air-launched radio guided missiles, known as Fritz X and the HS 293. The former was a free-fall bomb and the latter was a small jet aircraft. They were directed primarily at Allied surface vessels. As a refinement this type of missile could have been equipped with photo-electric, television, radar, or heat-seeking equipment for greater accuracy.

Although these missiles were primarily guided bombs, and were tracked optically from the launching craft, they foreshadowed the modern air-to-surface missile. With the current progress being made in defenses against manned aircraft attacks, it is now important to be able to launch missiles from the bombers while they are at some distance from their targets. The planes can then take evasive action, and the missiles can be guided onto their targets through heavy surface-to-air and air-to-air missile defenses.

THE EVOLUTION OF GUIDANCE TECHNIQUES

Radar Control

During World War II the Battle of Britain saw the first practical use of radar to detect enemy aircraft. It also directed fighter aircraft into position for attacks upon these airborne intruders. Although the control was not automatic and aircraft (instead of missiles) was guided into position, it contained some of the basic elements for remote radar guidance. Actually, the human pilot of the fighter aircraft provided the airborne control element which might have been replaced by an automatically controlled means for homing on the incoming bombers.

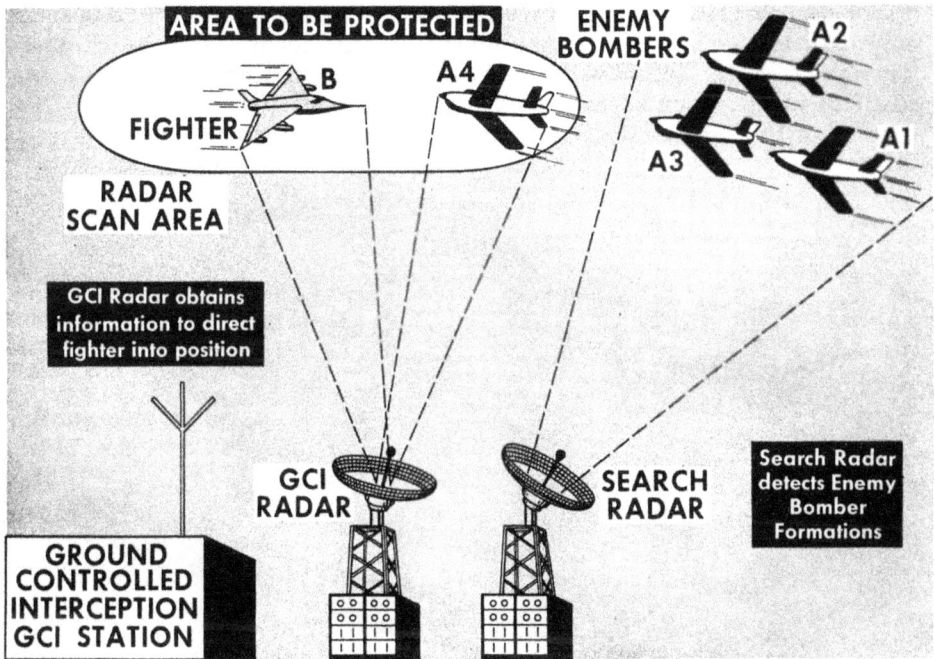

The search radar detected the incoming enemy bombers A_1, A_2, etc., and gave warning information to the GCI (ground-controlled interception) stations. At this location the GCI radar was brought into operation to scan an area to be protected from enemy bombers. A ground radar display showing both the fighter B and the bomber A_4 gave the control operator sufficient information to send a radio command signal to the fighter pilot. This signal directed him into the proper position to attack the oncoming bomber. Later developments included an airborne radar in the fighters, permitting pilots to obtain their own radar presentation for a very direct night fighter approach to the bombers. If the pilot could have been replaced by an automatic control system for guiding the fighter on the basis of the radar signals, a form of guided missile could have been placed in operation.

Radar Fire Control

Another World War II development was radar fire control. In such an operation the radar equipment followed the incoming bomber, and through a computer directed the antiaircraft gun into the proper position to fire into a zone calculated to intercept the bomber. Actually, the tracking movement of the radar antenna was fed across to the gun mount with corrections made by the computer for the proper lead angle. The arrangement used for transferring this motion to the gun was known as a *synchro system*. This technique was applied to gun direction in a number of designs for ground-based, shipborne, and airborne use during World War II.

From the development of components and circuitry for such closed-loop servo systems has sprung the muscular elements for modern guided missile control. Also, the accuracy demands on the radar equipment for fire control resulted in improved narrow-beam patterns for more precise tracking, thus providing the servomechanism with the data necessary for correct gun positioning.

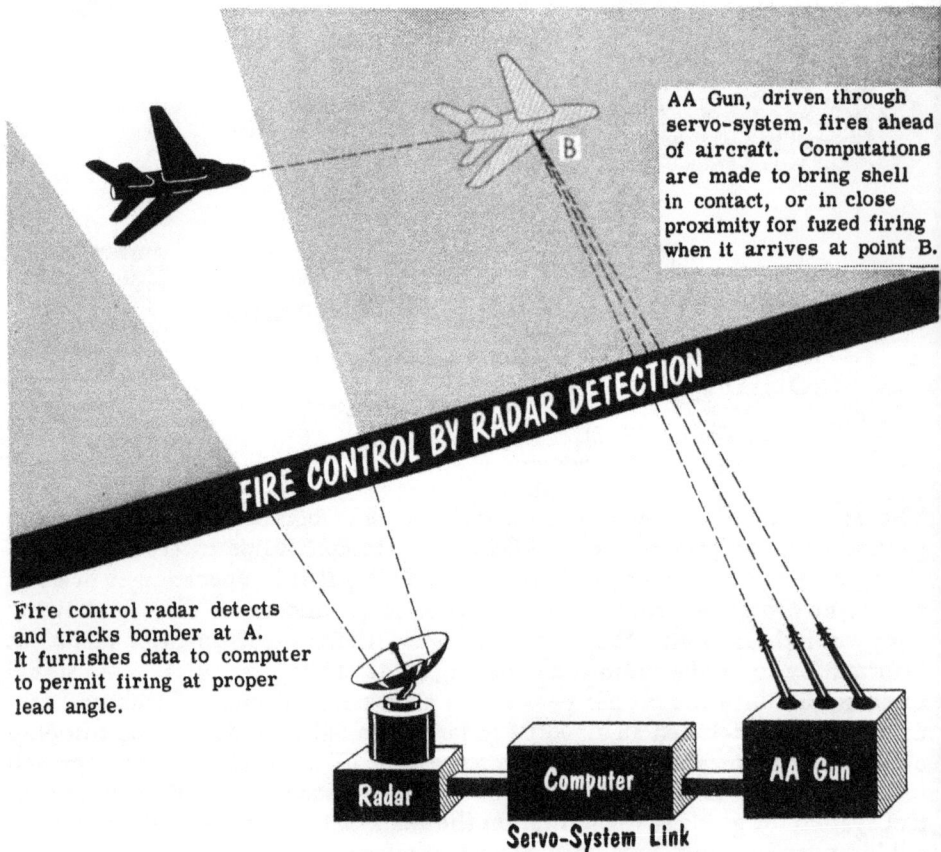

AA Gun, driven through servo-system, fires ahead of aircraft. Computations are made to bring shell in contact, or in close proximity for fuzed firing when it arrives at point B.

FIRE CONTROL BY RADAR DETECTION

Fire control radar detects and tracks bomber at A. It furnishes data to computer to permit firing at proper lead angle.

Radar   Computer   AA Gun

Servo-System Link

Radar Guidance

The above techniques of radar-controlled interception and fire-control provided a background for directing short-range missiles against moving targets, such as hostile aircraft. Radar fire-control techniques in particular contribute to a typical ground-launched missile system utilizing radar command guidance.

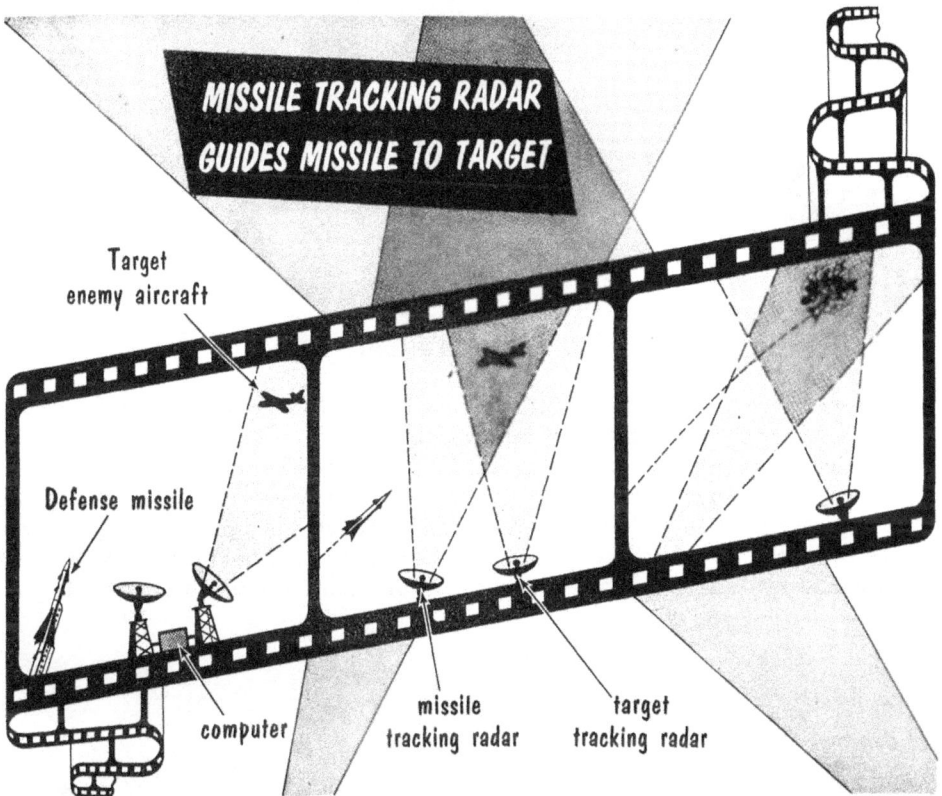

MISSILE TRACKING RADAR GUIDES MISSILE TO TARGET

Target enemy aircraft

Defense missile

computer

missile tracking radar

target tracking radar

Using two radars and a computer, both the target and the missile can be followed and the correct course for the missile determined. One radar tracks the target; the other the missile. Commands are sent to the missile over its radar guidance channel or by a separate communications channel. This basic technique, employed by the Germans during World War II, has since been refined to provide the requirements for successful guidance of ground-launched missiles against supersonic aircraft. The missile can be equipped with a proximity fuze to enable it to be effective even if it only manages to come within the vicinity of its target. However, such fuzes did not actually guide the shell and did not represent a form of missile guidance, although many smaller missiles are often so equipped.

Radar Guidance (contd.)

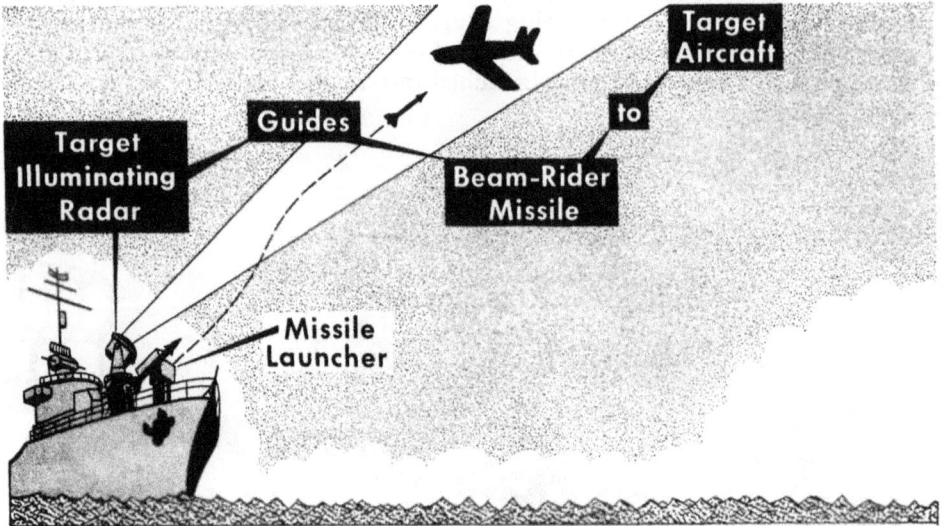

Another well known radar-guidance method is the *beam-rider* technique in which the target is *illuminated* by the ground radar. The missile is launched into this beam and follows it to the target. Deviations from the beam pattern are detected by a receiver in the missile. Corrections are made to insure that the correct path is followed. The beam-rider system is particularly attractive for those launching sites which do not have the large areas necessary for command systems. Shipboard and airborne launchers fit this pattern.

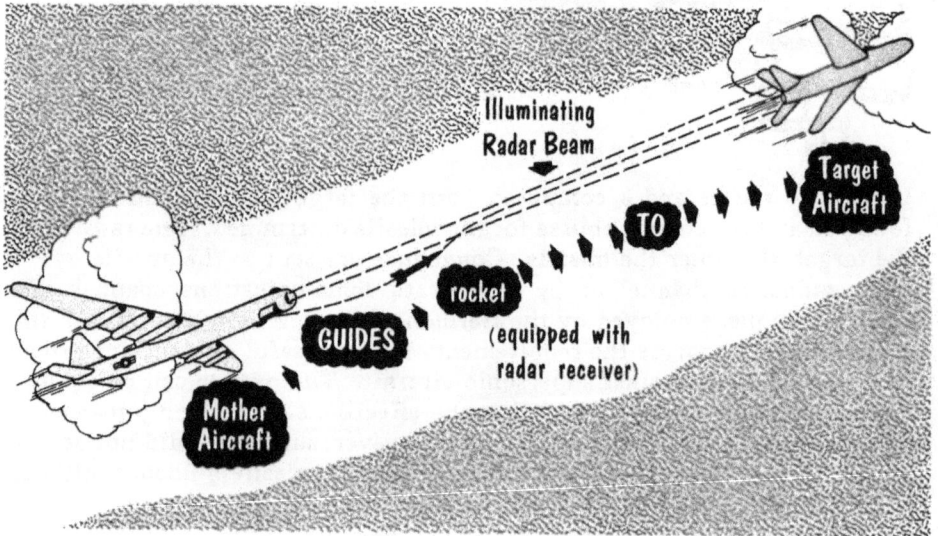

Homing by Radar and Infrared

Guided rockets launched from a mother aircraft may employ a self-contained radar receiver necessary to receive the radar reflections from a target *illuminated* by the launching aircraft. The rocket may operate as a beam rider and follow the illuminating beam to the target, or it may simply home on the reflected radar signal. This technique is known as *semi-active* homing, because although there is a radar signal from the launcher or a separate site the missile does not radiate a signal. Another form of target seeking, known as *passive* homing, is also utilized in some designs. In this type of guidance a receiver in the missile may pick up radar signals from the target aircraft, or an infrared receiver may utilize the heat radiations from the engines of the enemy aircraft.

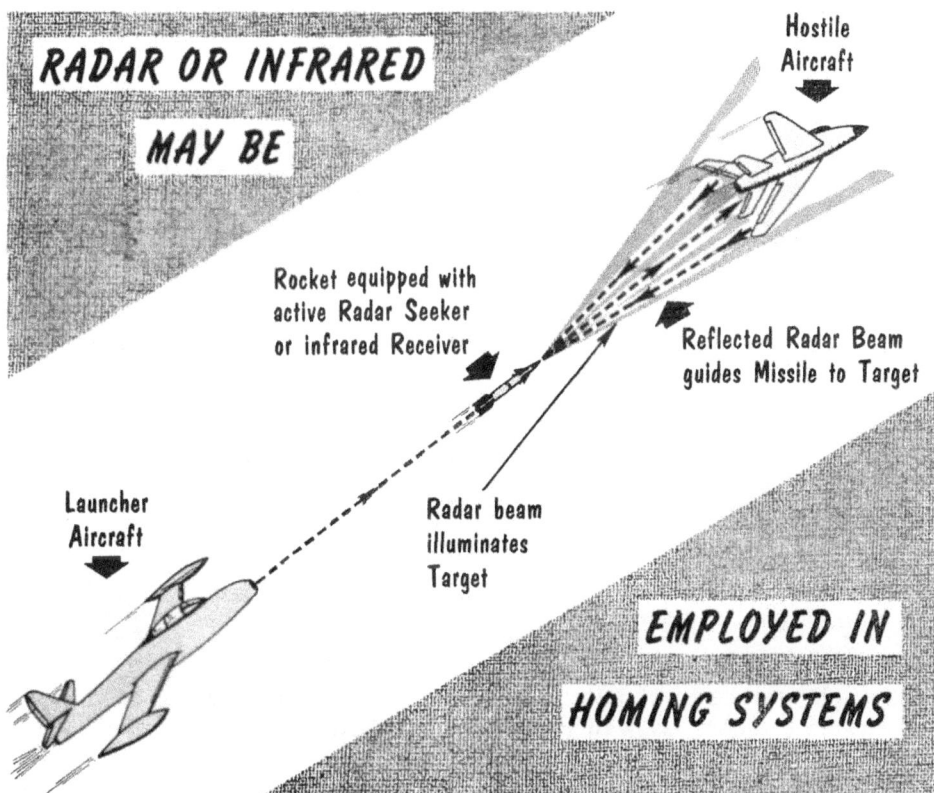

RADAR OR INFRARED MAY BE EMPLOYED IN HOMING SYSTEMS

Hostile Aircraft

Rocket equipped with active Radar Seeker or infrared Receiver

Reflected Radar Beam guides Missile to Target

Launcher Aircraft

Radar beam illuminates Target

In another form of guidance the air-to-air missile may contain a complete radar system within itself. The transmitter will radiate a signal to *illuminate* the target, and the reflected signals as received in the missile will provide the means for homing to it. This form of guidance is known as *active* homing because the missile is completely active in its seeking system.

THE EVOLUTION OF GUIDANCE TECHNIQUES

Navigation and Stabilization

Now that we have had a general look at the techniques employed in short-range missile guidance we can proceed with the study of techniques more closely related to long-range missile guidance. It was mentioned previously that a high order of navigational accuracy is required to insure that the warhead will be delivered within an acceptable radius of the target area. These missiles may utilize gyroscopes, star tracking, radar, and accelerometers to guide them to their destination. Gyroscopes are essential to almost all types of short- and long-range missiles to insure flight stability. But in the long-range missiles they may also contribute to or play a large role in the guidance as well. Star tracking, radar, and accelerometers are used as required, depending upon the particular guidance system selected for a given type of missile. The general concepts of long-range navigation were developed some time ago for ship and aircraft use. They are now applied to pilotless aircraft and rockets travelling at much higher speeds in the upper atmosphere or in space.

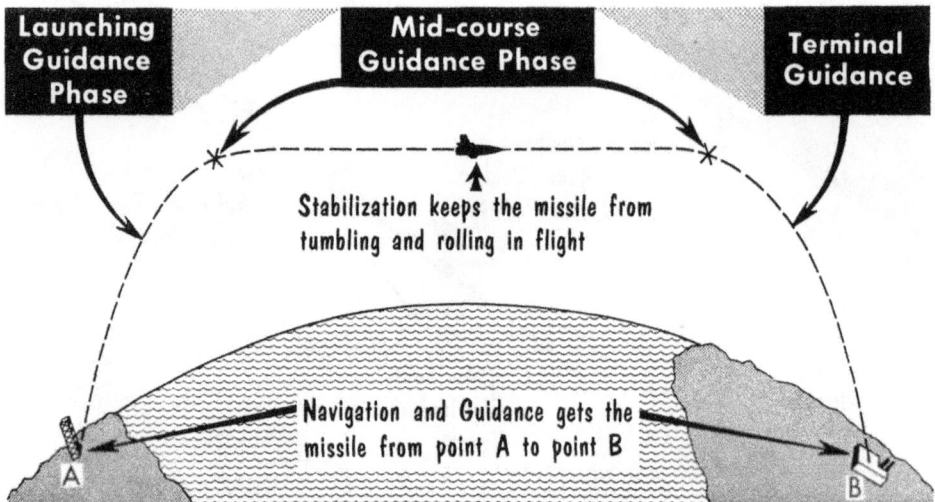

Launching Guidance Phase

Mid-course Guidance Phase

Terminal Guidance

Stabilization keeps the missile from tumbling and rolling in flight

Navigation and Guidance gets the missile from point A to point B

Although the phases of launching, mid-course and terminal guidance, are applicable to short- as well as long-range missiles, for the longer ranges these phases may come at considerable distances from the launching site and pose much greater problems on the overall system. One method for providing mid-course guidance for long-range missiles is to utilize the navigational information provided by the stars, somewhat as it is utilized by ships at sea. Due to the high speed of the missile, when celestial navigation is used, automatic "star-tracking" must be employed. However, as we shall see later, sufficient guidance is given in the launching or initial phase of some ballistic missiles to eliminate mid-course and terminal guidance.

Navigation and Stabilization (contd.)

In the early part of the 20th century the gyroscope was developed to the point where it could be utilized as a compass on shipboard. When pointed to the north and properly corrected, it gave better accuracy than the magnetic compass which was always susceptible to stray magnetic fields. Known as the *gyro-compass*, it could deliver signals to an automatic pilot to replace the helmsman for much of the steering effort. This application was called the *gyro-pilot*.

In another somewhat different application the gyroscope was developed to provide stabilization of the vessel in rough seas. Although initially very large gyros were used for this purpose, later forms utilized smaller gyros which worked through control-circuits to adjust fins below the water line to control the attitude of the vessel much as an autopilot functions in an airborne vehicle today.

The modern cruise-type long-range missile utilizes the gyroscope for stabilization and guidance. Sometimes, it also uses star-tracking for navigation. In addition radar may be used in the launching phase. Thus, three concepts of long-range navigation, which were conceived before the modern age of high-speed jet aircraft and rockets, may be utilized in an advanced form to guide missiles over long distances.

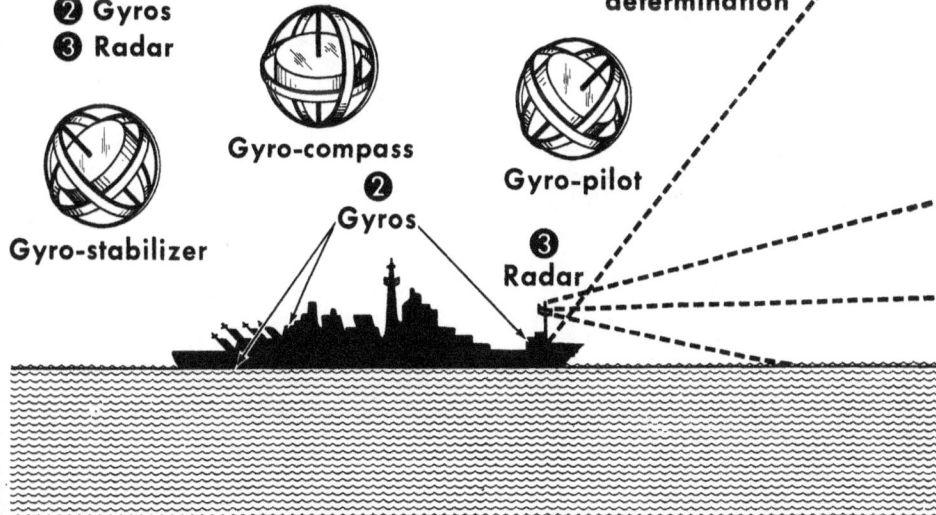

Navigation and Stabilization

Three Concepts of Long Range Navigation include:
1. Stars
2. Gyros
3. Radar

Gyro-stabilizer — Gyro-compass — Gyros — Gyro-pilot — Radar

1. Stars provide space references for position determination

Navigation and Stabilization (contd.)

Later in the 20th century the gyroscope, in a more compact form, was employed to detect deviations in an airplane's flight attitude. This application of the gyroscope closely paralleled that of the ship's gyro-pilot, and permitted the human pilot to relinquish control without sacrificing flight stability. The autopilot is a fundamental requirement for practically all missiles to insure flight stability. The guidance system will feed in signals to control the autopilot. The basic stability provided is essential to exercise effective control.

Both the ship and the airplane adopted radar for navigational purposes almost simultaneously in World War II. Modern airborne-radar systems can indicate the velocity of the craft with respect to the earth's surface, and thus provide important data for determining the distance travelled along the path.

THE AUTOPILOT IS A FUNDAMENTAL REQUIREMENT FOR ALL MISSILES TO INSURE STABILITY OF FLIGHT

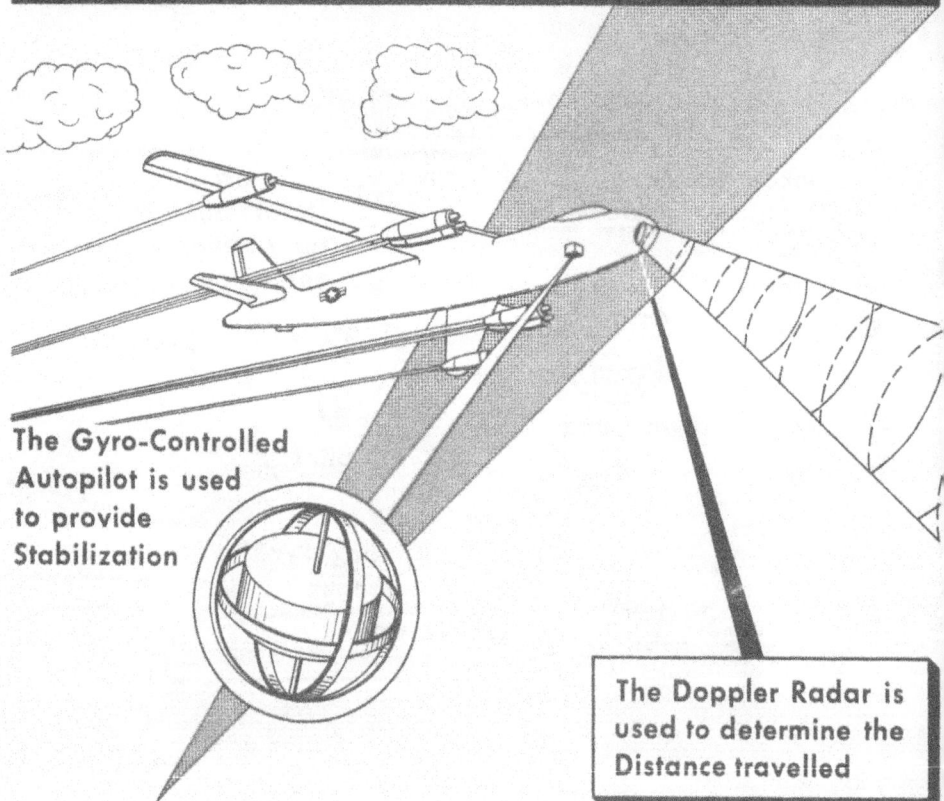

The Gyro-Controlled Autopilot is used to provide Stabilization

The Doppler Radar is used to determine the Distance travelled

Cruise and Ballistic Operations

The V-1 missile is a good example of a cruise-type missile with a range slightly over 200 miles. The range was so short that gyro stabilization and fuel cutoff could be used to control its trajectory. The concepts of long-range missiles were not required. The principles of the autopilot were fully utilized in this application. Modern long-range unmanned jet aircraft, such as the Snark, require the concepts of long-range navigation, and must utilize the gyroscope to the fullest extent of its capabilities as well as star-tracking and possibly radar to maintain a guided path across thousands of miles of land and sea.

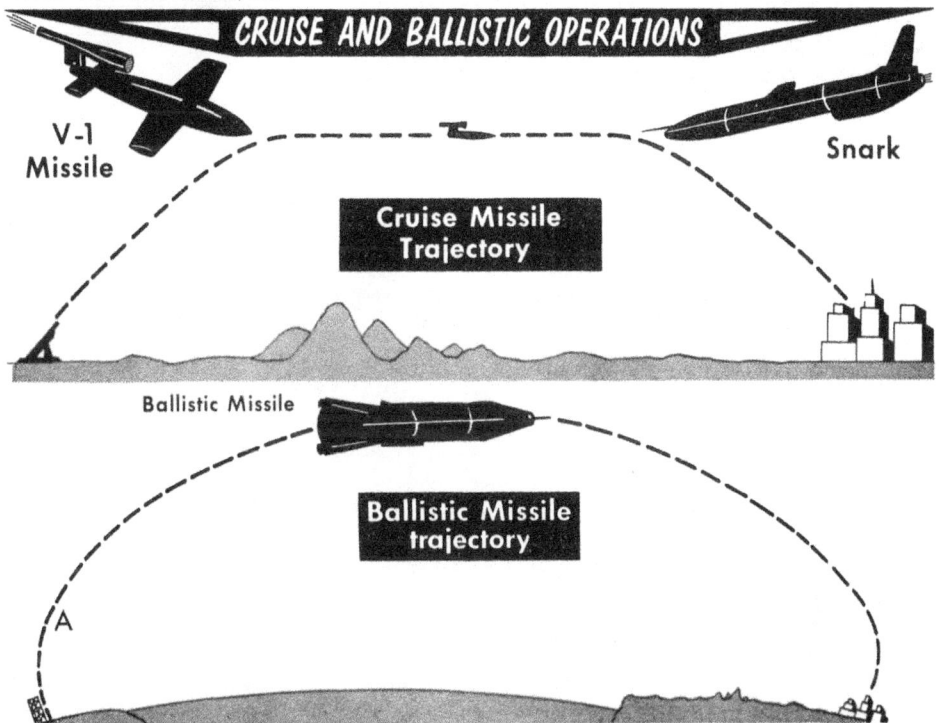

CRUISE AND BALLISTIC OPERATIONS

V-1 Missile

Snark

Cruise Missile Trajectory

Ballistic Missile

Ballistic Missile trajectory

A

Gun-fired shells travel in a ballistic trajectory. Much of the shell's accuracy depends upon the ability of the gun to fire it into the proper trajectory. The rifling of the gun barrel gives it a spin which produces a gyroscopic effect imparting stabilization.

The modern ballistic missile, whether short- or long-range follows the same type of trajectory. It must also be aimed correctly to arrive at its desired destination with an acceptable degree of accuracy. The rocket must be controlled during a portion of its flight shown at A to aim it into its trajectory in such a manner that it will travel to the desired target area with the required accuracy. This aiming is done by gyroscopic or accelerometer control, and in some cases with radar assist.

THE EVOLUTION OF GUIDANCE TECHNIQUES

Self-Contained Guidance

Completely self-contained guidance of a long-range missile has been the objective of workers in this field for some time. Although not representative of a modern guided missile system, the first application of a self-contained system dates from the latter part of the 19th century when gyroscopic control was applied to naval torpedoes for short-range underwater operation. In this case the steering rudders of the torpedo were controlled to bring it back on course when it deviated from a gyro-stabilized frame of reference.

EXAMPLE OF SELF-CONTAINED GUIDANCE

A gyroscopic device contained in the torpedo maintains the direction of the path to the target

In the modern long-range ballistic missile, guidance can be achieved by a combination of gyroscopes, accelerometers, associated computer, and control elements. In a technique known as *inertial* navigation, the gyros provide a stable platform of reference. The accelerometers detect motions away from a desired course. The computer calculates the necessary corrections to be applied for returning the missile to the required trajectory. Actually, this control need not be exercised throughout the entire flight of the missile, but can be applied effectively if used only during an initial portion of the trajectory. The missile can be placed into the proper ballistic trajectory by this procedure just as the gun in a previous example fired a shell into the proper trajectory to control accurate artillery fire.

Summary

In summary, it can be said that missiles fit well into the classification of moving-target and fixed-target types. Moving-target types have been well developed for defense against high-speed jet bombers. In the future, they will undoubtedly be required to intercept missiles moving at much higher speeds, such as the intermediate range and intercontinental ballistic types. The fixed-target missiles find their greatest challenge in achieving high accuracy against targets at extreme ranges. Such performance in an operational sense of highly practical form is yet to be achieved for some types. An equally severe test is encountered in placing satellites into precise orbits about the earth.

Gyroscopes are seen to be essential to practically all types of missiles for stabilization purposes. *Radar* and *infrared* techniques are of key importance in moving-target types *Precision gyroscopes* and *accelerometers* are the key elements of long-range fixed target missiles. The accurate mathematics of *airborne* and/or *ground computers* is necessary for many types. The capabilities of *analog* and *digital computers* are applied to fit the particular requirement. *Servomechanisms* make possible the exercise of control after the data is received. All of these tools will require extensive development to insure the proper control of satellites and space vehicles.

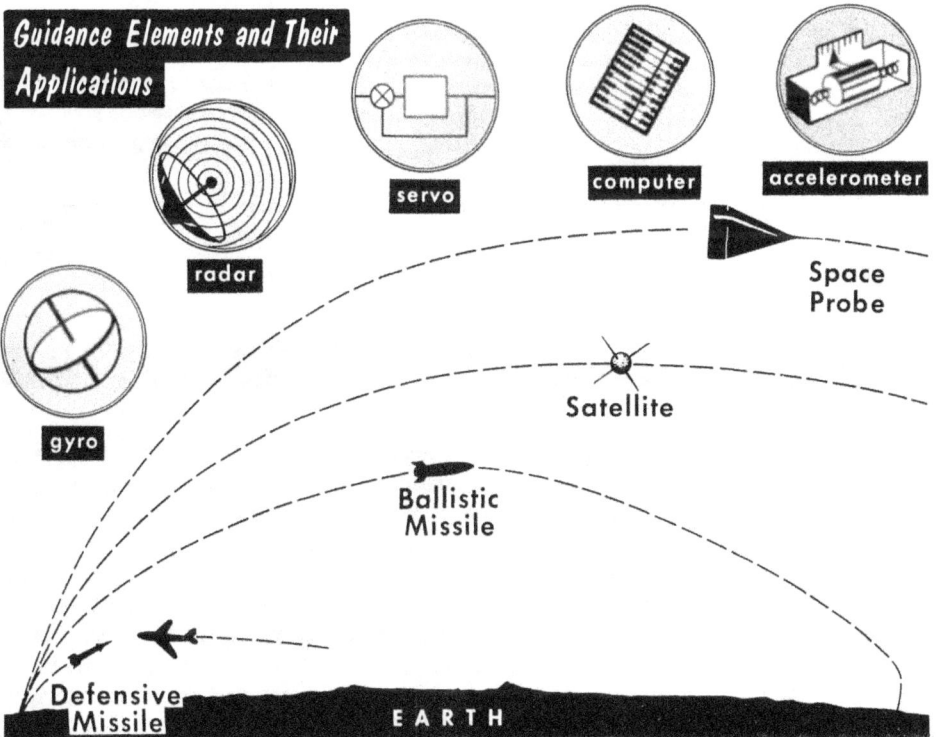

Guidance Elements and Their Applications

servo

computer

accelerometer

radar

Space Probe

gyro

Satellite

Ballistic Missile

Defensive Missile

EARTH

Automatic Control

Simple control systems in which an input signal controls an output element are commonplace in modern man's environment. The flick of a light switch controls the amount of illumination in a room. Pushing a button brings an unattended elevator to the correct destination. The voltage applied to a meter causes an indicator to move across a scale. However, these examples are of the simplest type of control — the open-loop system.

Control systems divide into two broad groups — open-loop and closed-loop types. In the first type, the output is controlled only by the input level. In the second type, a portion of the output is fed back to the input to achieve a coordinated regulation between the response of the controlled element and the controlling input signal. Although in missile stabilization and guidance we shall be concerned exclusively with closed-loop feedback systems, it is well to compare the two types to show the advantages gained by applying feedback.

These are Examples of Automatic Control Systems:

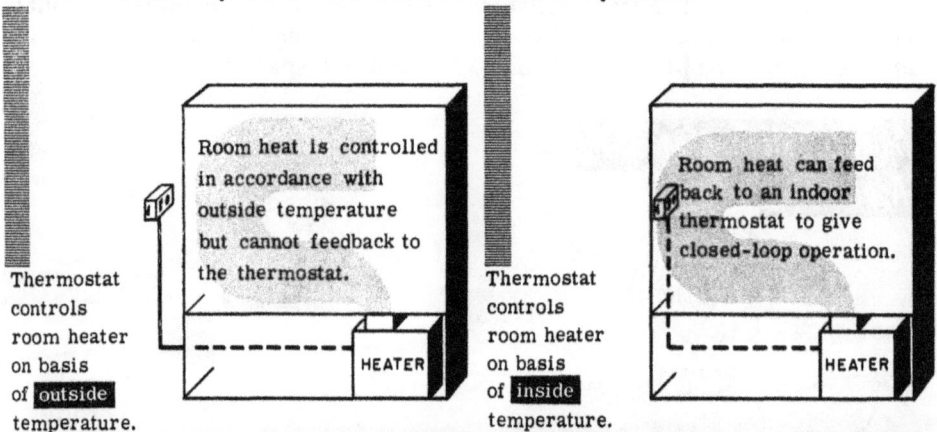

Room heat is controlled in accordance with outside temperature but cannot feedback to the thermostat.

Thermostat controls room heater on basis of outside temperature.

Room heat can feed back to an indoor thermostat to give closed-loop operation.

Thermostat controls room heater on basis of inside temperature.

(A) Open-Loop Control   (B) Closed-Loop Control

The difference between open-loop and closed-loop systems can be illustrated by comparing two arrangements of thermostatic heat regulation. In type A a thermostat is placed outside of the room. And while it responds readily to outdoor temperature changes it cannot regulate the heat as well as type B, where some of the heat in the room feeds back to the thermostat. In the latter case as the room is heated above the desired temperature, the heater is regulated to a lower level and the temperature is brought down to a level a little below the desired amount. Then the thermostat acts again to bring the heat up slightly and the temperature stabilizes around the desired value if the control system has enough sensitivity and high enough feedback.

Feedback Diagrams and Equations

The relationship between input, sensitivity, output, and feedback is shown by the following diagrams and equations:

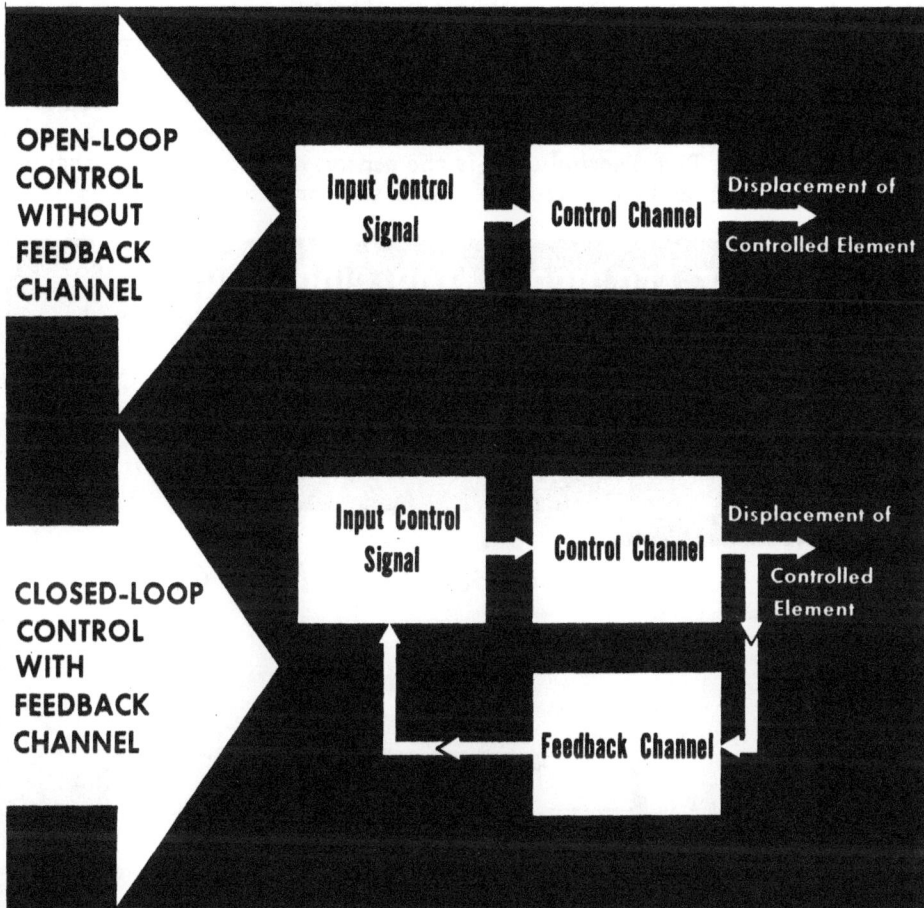

If the input is designated as R and the transfer gain of the control channel as G, the displacement (D) for an open-loop system may be expressed as $D = G \times R$. With the same designations and with a feedback level F the displacement for a closed loop-system is expressed as $D = GR/(1 + GF)$. If the gain is high in the control channel and if the feedback factor is such that the product of $G \times F$ is substantially greater than unity, $D = R/F$ or $D/R = 1/F$. Thus, the displacement becomes independent of the control channel in a sense and the gain is dependent upon the feedback alone. This condition results in high stability, greater freedom from drift, and an improvement in the response speed.

Example of a Closed-Loop System

An example of a closed-loop system, very familiar to the radio industry, is the automatic frequency control (afc) of a superheterodyne oscillator. In an FM receiver this stabilization is brought about by returning some of the discriminator d-c voltage as a bias to the grid of a reactance-control tube connected to the local oscillator. When properly phased the voltage through the control channel will cause the oscillator to return to a frequency which produces zero-error voltage at the discriminator. A typical closed-loop error-correcting system is formed. The oscillator frequency is the controlled situation. The discriminator is the sensor. The d-c voltage channel is the control link, and the reactance tube is the corrector.

THE AFC OSCILLATOR CIRCUIT OF A SUPERHETERODYNE RECEIVER IS AN EXAMPLE OF A CLOSED-LOOP SYSTEM

Closed-Loop Error Corrector

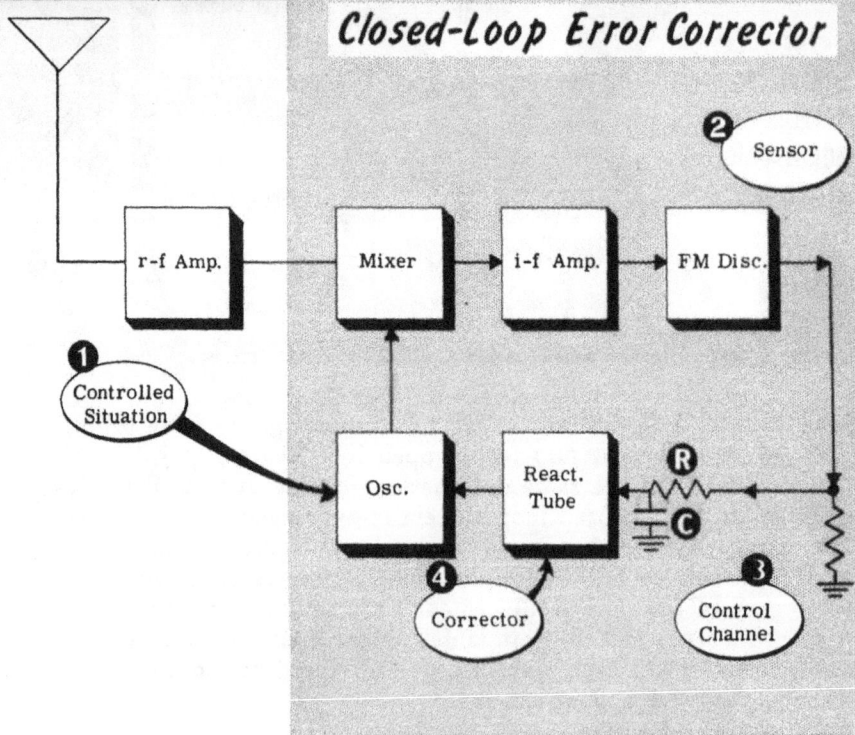

Flight Control as a Closed-Loop System

Although man has great capability to respond to the requirements for a flexible control link, many high-speed devices can be controlled more effectively by an automatic means. A good example can be shown by one phase of automatic flight control.

The pilot of an aircraft observes that the attitude of the plane is changing from the desired position. He wishes to stabilize it. He senses the *off-course* condition, and decides how much force to apply to the control channel to achieve correction. This control is applied to the rudder or elevators, and the plane resumes the desired attitude. In this example, man is the *sensing* element as well as part of the closed-loop system functioning also as a computer.

The Pilot Is The Sensing Element As Well As The Computer in A Closed-loop Control System

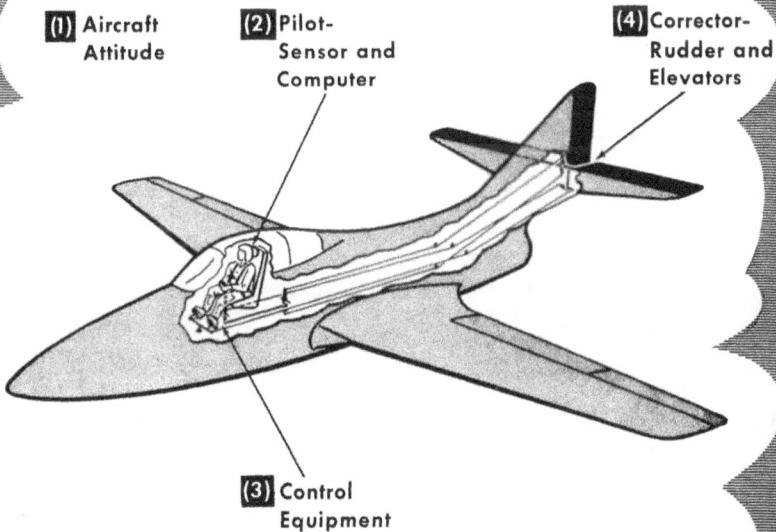

(1) Aircraft Attitude

(2) Pilot-Sensor and Computer

(4) Corrector-Rudder and Elevators

(3) Control Equipment

The Elements of Automatic Control

THE ELEMENTS OF AUTOMATIC CONTROL include

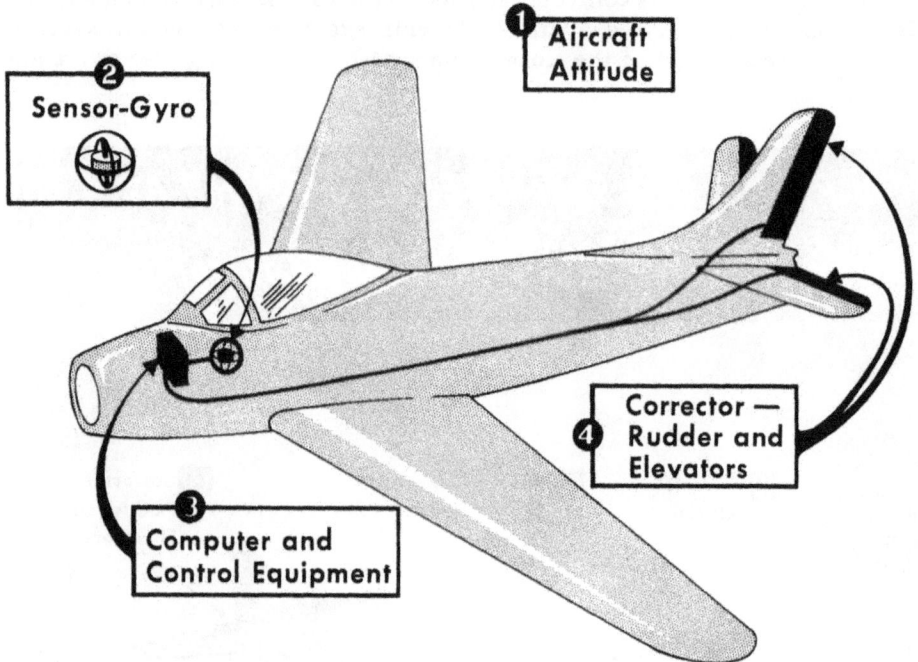

1 Aircraft Attitude

2 Sensor-Gyro

4 Corrector — Rudder and Elevators

3 Computer and Control Equipment

The pilot can be replaced in so far as the *sensing* and *computing* elements are concerned by installing a suitable gyroscope and computer combination in the aircraft. The gyro will sense the deviation of the plane's attitude. Signals will be fed to the computer which will provide the necessary information to operate the control channel and provide the proper action to the plane's control surfaces. The desired attitude of flight will be restored just as it was in the manned operation. Thus, the gyro and computer replace man and serve the purpose of sensing and computing in the closed-loop system.

The Elements of a Modern Control System

Separation of sensing and computing is generally essential in modern control systems because in many instances the sensing device cannot perform the computer function. There is also need for a *reference* against which the computer can make comparisons to determine deviations of the desired position of the craft. A separation of these functions of an automatic control system leads us to a closed-loop diagram with a greater number of elements.

THE ELEMENTS OF A MODERN CONTROL SYSTEM FORM A CLOSED-LOOP

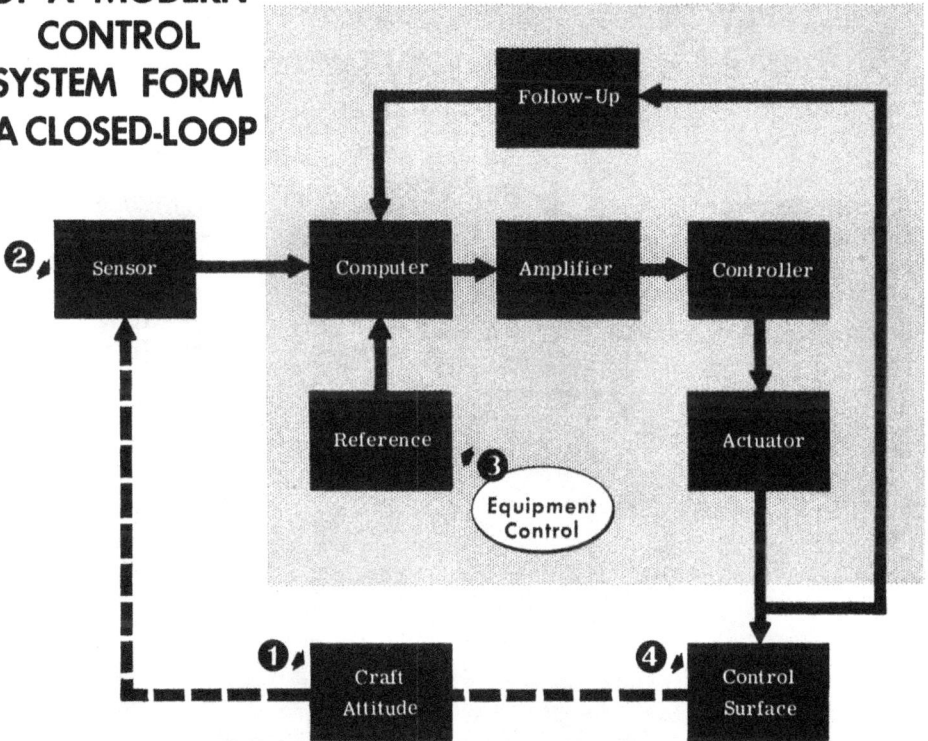

As information is taken from the computer, we find that it is necessary to amplify it before applying it to the controller. The controller must be followed by some type of actuator — mechanical, electrical, hydraulic, or pneumatic — to move the control surfaces. If we overcontrol, the aircraft will execute violent movements rather than smooth corrections in its attitude. Some form of feedback is desirable to keep a smooth corrective action. Thus, we arrive at eight divisions of the automatic control loop. All of these portions will be dealt with in greater detail later. For the moment it is sufficient to illustrate these functions by familiar analogies.

(1-21)

Man as a Control System

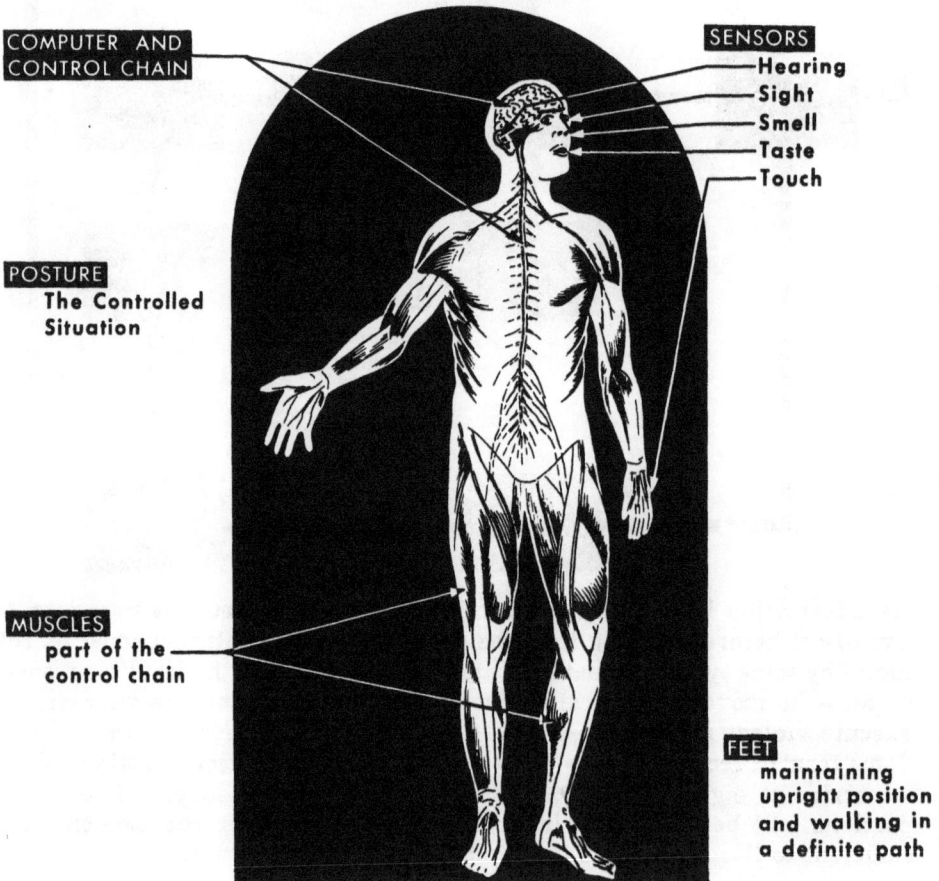

It appears that man himself is one of the best analogies of the control systems which he is building to substitute for his role in modern craft. From the time of learning to walk a child exercises a number of elements which closely parallel those of the basic closed loop previously described. Five sensing channels, sight, hearing, smell, taste, and touch, feed inputs to the brain. With its marvelous capacity as a computer, the brain provides direction to the muscular system. In turn the arms and legs are set into motion commensurate with required action.

MAN AS A CONTROL SYSTEM

COMPUTER AND CONTROL CHAIN

SENSORS
Hearing
Sight
Smell
Taste
Touch

POSTURE
The Controlled Situation

MUSCLES
part of the control chain

FEET
maintaining upright position and walking in a definite path

The Flight Control Surfaces

In going from the controlled motions of man to those of an airplane, we find that three directions of rotation must be influenced in the control of flight attitude. They are about the longitudinal, lateral, and vertical axes of the craft, and are known as roll, pitch, and yaw. When these rotations are properly controlled the plane will assume a stabilized attitude of flight.

Since the plane is flying in air its rotational positions may be controlled by elements which affect the flow of air about its fuselage, wing, and tail structure. A typical set of control surfaces are the elevators, rudder, and ailerons. The outputs of a flight control system may drive the ailerons, elevators, and/or rudder through electrical motors, hydraulic or pneumatic actuators to place them in the correct position for the proper flight path.

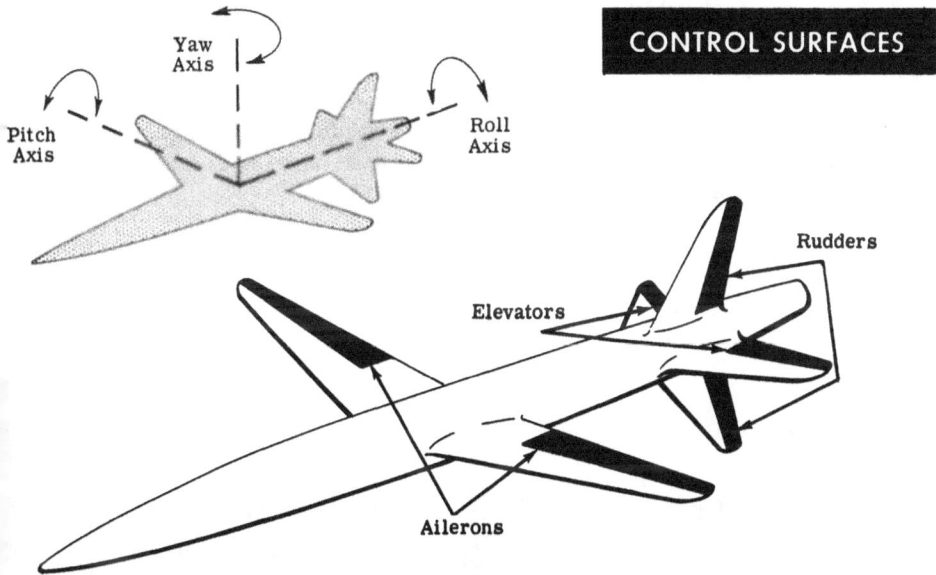

CONTROL SURFACES

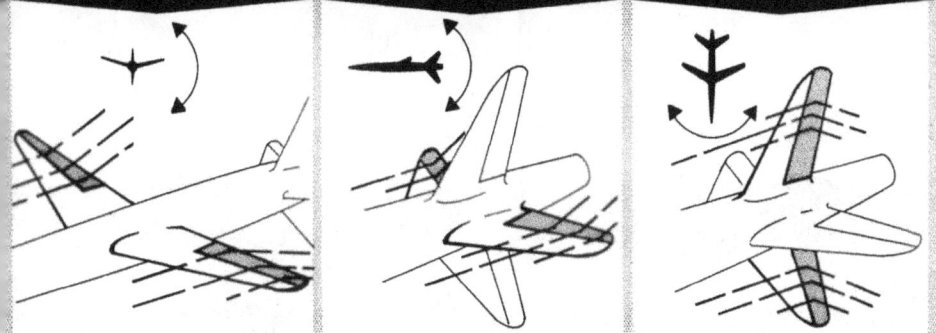

| Roll is influenced by the position of the ailerons. | Pitch is influenced by the elevators. | Yaw is influenced by the position of the rudder. |

Rocket Thrust Control and Direction

While several guided missiles of the past have been in the pilotless aircraft class, and some types in operation at present are really jet aircraft, the trend has been toward the use of rockets for both short- and long-range applications for a number of years. One of the principal difficulties in stabilizing a rocket is brought about by the shift in its center of gravity as the fuel is consumed. This situation is even more significant for large rockets, where the ratio of fueled weight to empty weight is high. The earliest artillery rockets, which were relatively small and light, achieved a degree of stability by employing a tail stick which gave a better weight equalization as the fuel was burned.

A later development utilized stabilizing fins located near the base of the rocket, and even today many missiles use a certain amount of fin stabilization. However, it should be recognized that these arrangements provide only inherent stability and do not permit any controlled corrections, unless portions of the fins are movable and adjustable by a remote control system. It is also important to bear in mind that fins or any other external aerodynamic surfaces are useful only while the missile is travelling through the atmosphere. Here the situation is much the same as it is with an aircraft where the tail structure acts to stabilize pitch and yaw.

Small Solid
Propellant Rocket

Liquid or Solid
Propellant Rocket

Stabilizing Stick is used
for weight equalization
as fuel is consumed.

Stabilizing Fins are used
only when missile is
traveling through atmosphere.

Rocket Thrust Control and Direction (contd.)

Another technique for stabilizing a rocket both in the atmosphere and outside is that of deflecting the thrust gases of the rocket so that a rotational effect is imparted to the body in flight. This effect may be achieved by vanes in the path of the rocket thrust or by oblique discharge nozzles. Small rockets can be stabilized by these techniques, but the larger rockets require more elaborate arrangements to insure that stabilization can be controlled in the presence of winds or due to variations in the performance of the missile.

The first long-range missile employing adjustable vanes in the exhaust stream of the rocket engine was the German V-2. A combination of movable tail-fin sections and jet vanes controlled the roll, pitch, and yaw of the missile when actuated under the influence of corrective signals from suitable gyroscopes rotating along the appropriate axes.

The tail fins N and S and jet vanes N and S moving in tandem controlled direction (together) and roll (in opposition); jet vanes E and W controlled pitch (acting as elevators) and tail fins E and W acted as stabilizers. Jet vanes E and W were controlled by one gyroscope, which in turn was controlled by a master gyroscope. All other vanes and fin sections were controlled by a third gyroscope.

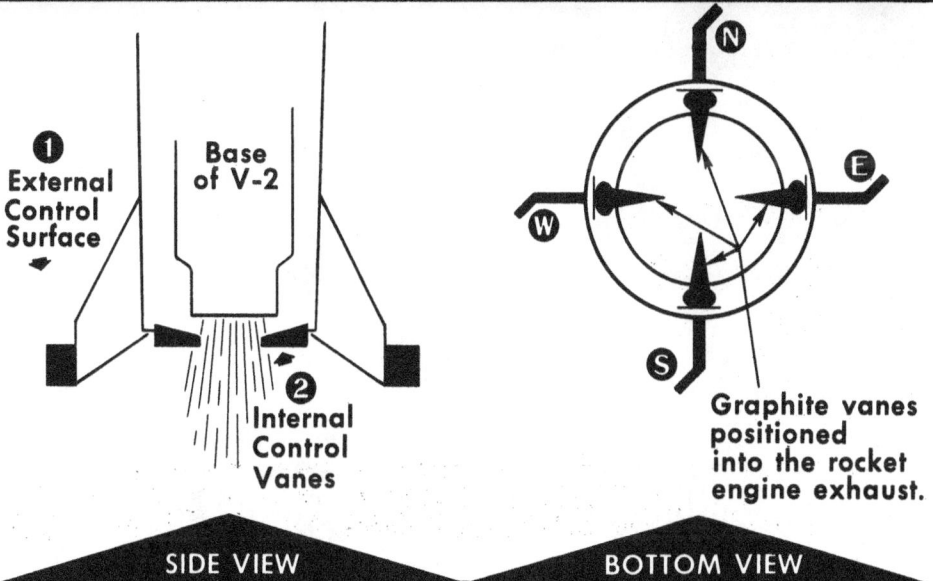

ROCKET THRUST CONTROL AND DIRECTION ARE CONTROLLED BY POSITIONS OF

❶ External Control Surface

Base of V-2

❷ Internal Control Vanes

N

E

W

S

Graphite vanes positioned into the rocket engine exhaust.

SIDE VIEW

BOTTOM VIEW

Rocket Thrust Control and Direction (contd.)

The second arrangement for controlling the flight attitude of a rocket is the gimbaled engine. In this design the exhaust stream can be directed at any desired angle to place the missile into the required path. Basically, this method is not greatly different from the variation of exhaust direction to obtain missile spin as mentioned above.

By controlling combinations of jet vanes and external fins, or by a gimbaled engine, rockets may be made to traverse vertical flight paths or curved trajectories as the application demands. A typical system for controlling the vertical stability of a rocket in flight and for placing it on any desired path deviating from the vertical is a combination of three gyroscopes, feeding signals through autopilot amplifiers to control roll, pitch, and yaw separately. If the gimbaled engine is employed, the control system will direct the thrust chamber to correct the pitch and yaw. Vernier jets or engines mounted along the sides of the rocket may be directed by the control system to regulate the roll. Vernier engines can also be employed to supplement pitch and yaw control as well as to provide accurate control of the cutoff velocity for ballistic rockets.

Rocket engine mounted in gimbals can be controlled for proper thrust direction to control pitch and yaw

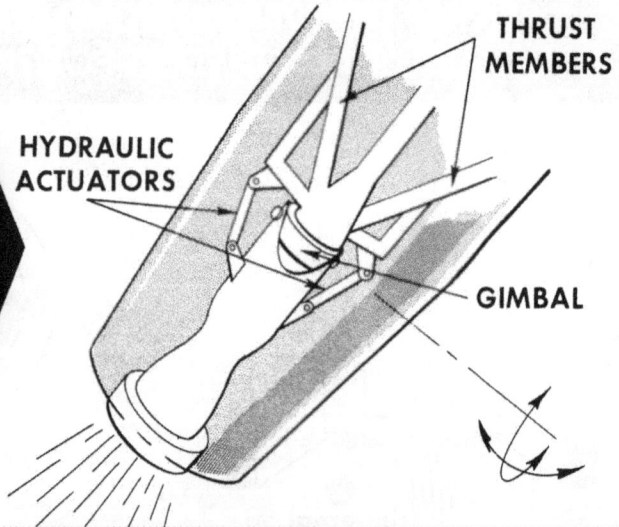

THRUST MEMBERS

HYDRAULIC ACTUATORS

GIMBAL

GIMBAL-MOUNTED FIRST STAGE ROCKET MOTOR

Rocket Thrust Control and Direction (contd.)

These small auxiliary thrust-directors are preferably mounted on fins or outrigger arms so that they can have a larger moment arms for control. At launching, the verniers are usually ignited first so that their control is in effect before the main engines are fired. Then they can control the vehicle from the moment of the lift-off from the launching pad. If the angle of the vernier engines must be varied sufficiently to direct their thrust even partially toward a part of the main rocket body, heat shields must be employed to protect it.

The rocket thrust-control discussed above applies to missiles using liquid propellants. For these missiles, valves can be controlled to shut off the fuel to the missile engine. When solid propellants are employed, the burning of the fuel cannot be stopped until the fuel supply is exhausted. Therefore reverse thrust is used as shown. The control of the direction of the thrust vector is somewhat similar to that of liquid-fueled engines. In one of the applications shown, the Jetavator swivels in and out of the thrust path causing it to change direction; in another application a flexible nozzle is coupled to the combustion chamber, and swings around like a gimbaled engine.

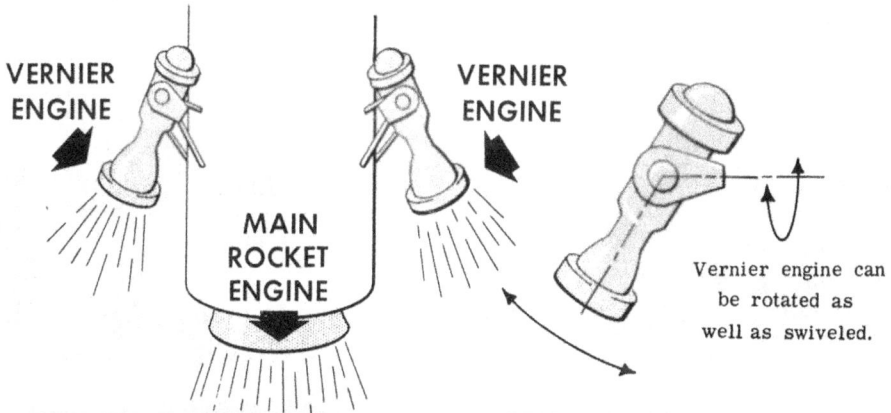

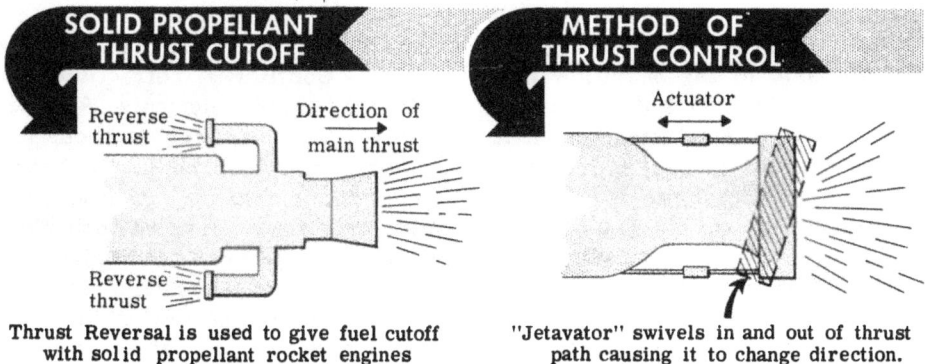

VERNIER ENGINE

VERNIER ENGINE

MAIN ROCKET ENGINE

Vernier engine can be rotated as well as swiveled.

SOLID PROPELLANT THRUST CUTOFF

Reverse thrust

Direction of main thrust

Reverse thrust

Thrust Reversal is used to give fuel cutoff with solid propellant rocket engines

METHOD OF THRUST CONTROL

Actuator

"Jetavator" swivels in and out of thrust path causing it to change direction.

Rocket Thrust Control and Direction (contd.)

If vertical flight is desired, the overall control or autopilot system will attempt to keep the missile stabilized within the capabilities of the gyroscopes and the remainder of the system. If a curved path of flight is desired, the pitch gyro must be precessed to impart the desired variation from the vertical. The gimbaled engine and the verniers will respond through the control system, and will align the missile heading with the one established by the gyro direction.

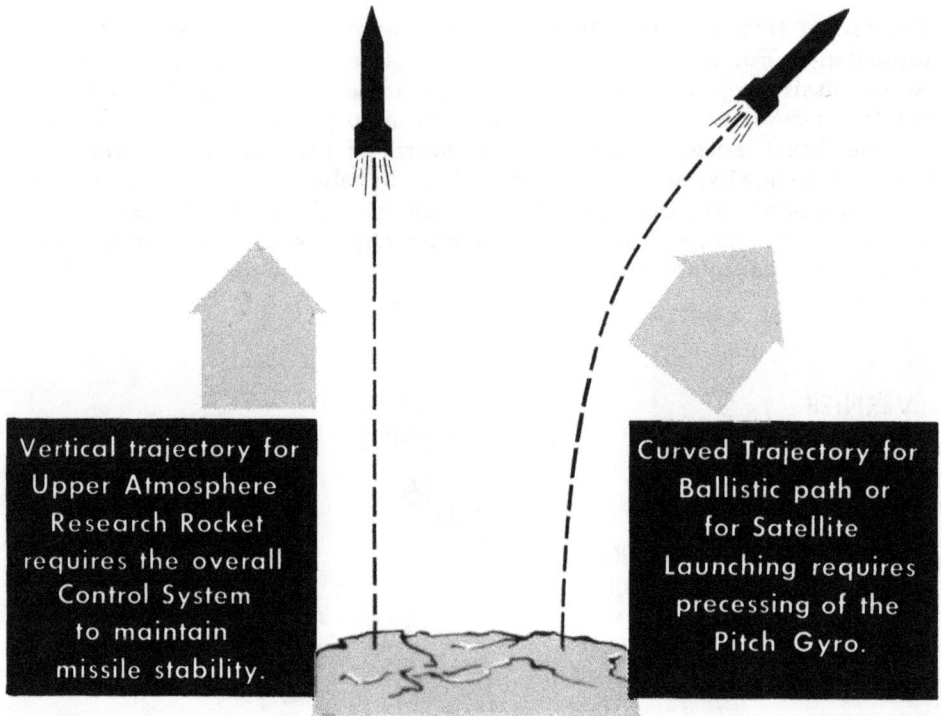

Vertical trajectory for Upper Atmosphere Research Rocket requires the overall Control System to maintain missile stability.

Curved Trajectory for Ballistic path or for Satellite Launching requires precessing of the Pitch Gyro.

In missile control systems proportional-type control is most widely used. In this type the correction for any deviation from a desired flight path is proportional to the deviation from the path. Continuous corrections for small deviations from the path are applied to the control elements. Another type of control system, known as the *on-off* or *bang-bang* method, has found only slight usage. In its operation the control elements are driven continuously from one direction to the other to a maximum extent and the length of time which they deliver output in one direction or the other determines the resultant accelerating force. Such a system can be simpler than the proportional type, but it has not found much appeal among missile-system designers. It is apparent that it would present a somewhat different feedback problem, and would consume more power to drive the system.

FUNDAMENTALS OF CONTROL AND GUIDANCE

Control and Guidance

The difference between *control* and *guidance* can be illustrated by considering the remote control of model aircraft in two phases. A model aircraft should be built in such a manner that stable flight is achieved to the necessary degree through its construction. However, if one desired to insure its stability a small gyro could be placed inside the model to control the ailerons, and thus prevent it from rolling.

If two gyros were used the pitch could also be controlled. Then the model aircraft would be controlled for flight stability in both roll and pitch. We would now have a model which would not pitch or roll within limits, but it would still be free to do maneuvers about its yaw axis. If the rudders were gyro-controlled too, the model would fly in the direction of original heading — if all went well. However, *guidance* of the model about the yaw axis can be brought into play by radio signals from the ground. This action would represent a simple form of command guidance. In this example, control and guidance are clearly separated. The craft is controlled about two axes by internal gyro-sensing and is guided along a third axis by radio signals from the ground.

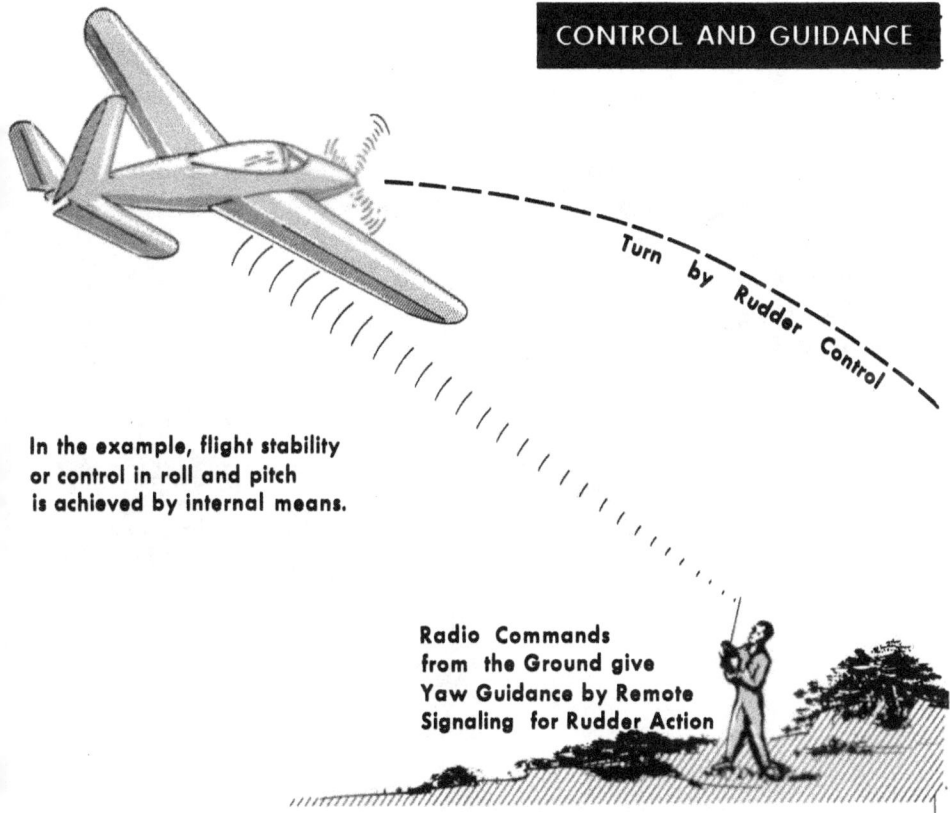

CONTROL AND GUIDANCE

Turn by Rudder Control

In the example, flight stability
or control in roll and pitch
is achieved by internal means.

Radio Commands
from the Ground give
Yaw Guidance by Remote
Signaling for Rudder Action

The Elements of Guidance

Guidance systems differ from the simpler single-loop control systems in that, generally, they involve multiloop feedback servomechanisms. Additional inputs beyond those required for stability alone are necessary to guide the missile to a target. For instance, in addition to complete closed-loop servosystems in the craft (to insure that it maintains the proper flight attitude) other sensing, computing, and control inputs will be required, to insure that a command guided surface-to-air missile is to seek out the incoming bomber. The radar guidance and the ground computer will form a loop completely outside of the missile, which may contain the remainder of the system.

Within the missile the guidance function is superimposed upon the autopilot system of the missile deviating it as required from a linear trajectory to reach the point in space where it can perform the mission for which it was designed. In the simplest types of guidance the deviations from an autopilot course may be supplied from a preprogrammed input. More elaborate types of guidance require that data relating to the actual trajectory be continually fed into the control system and referenced to the target or to a precalculated course.

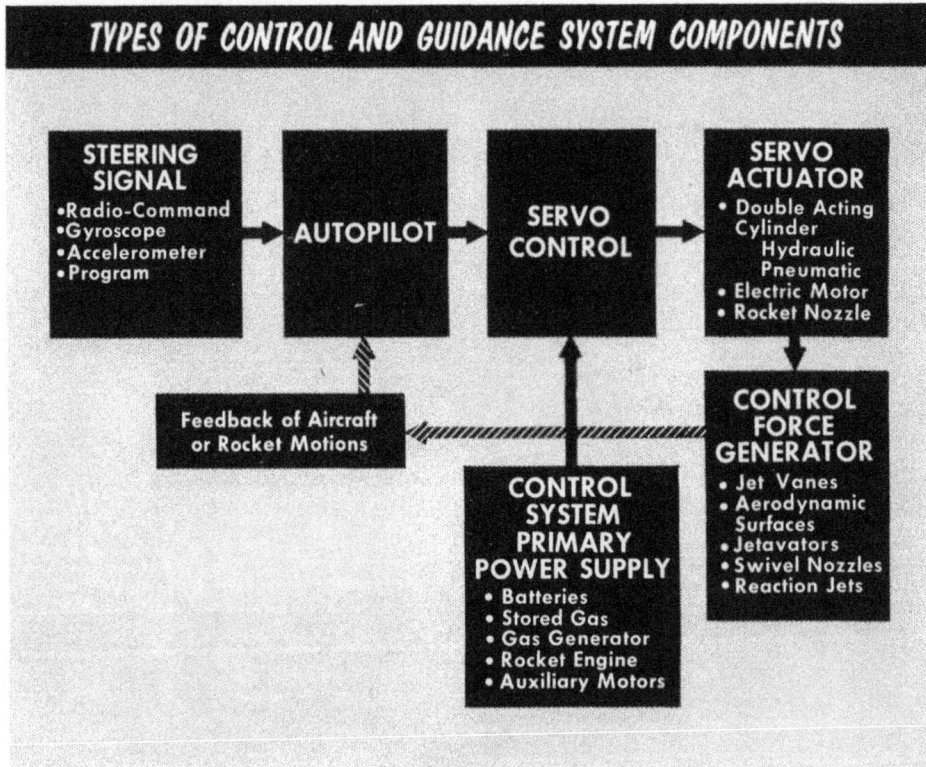

TYPES OF CONTROL AND GUIDANCE SYSTEM COMPONENTS

The Elements of Guidance (contd.)

THE COMPUTER, OR BRAIN, RESPONDS TO
INTERNAL SIGNALS, AND CORRECTIVE ACTION IS TAKEN
THROUGH THE MUSCLES AS A SERVO CHANNEL.

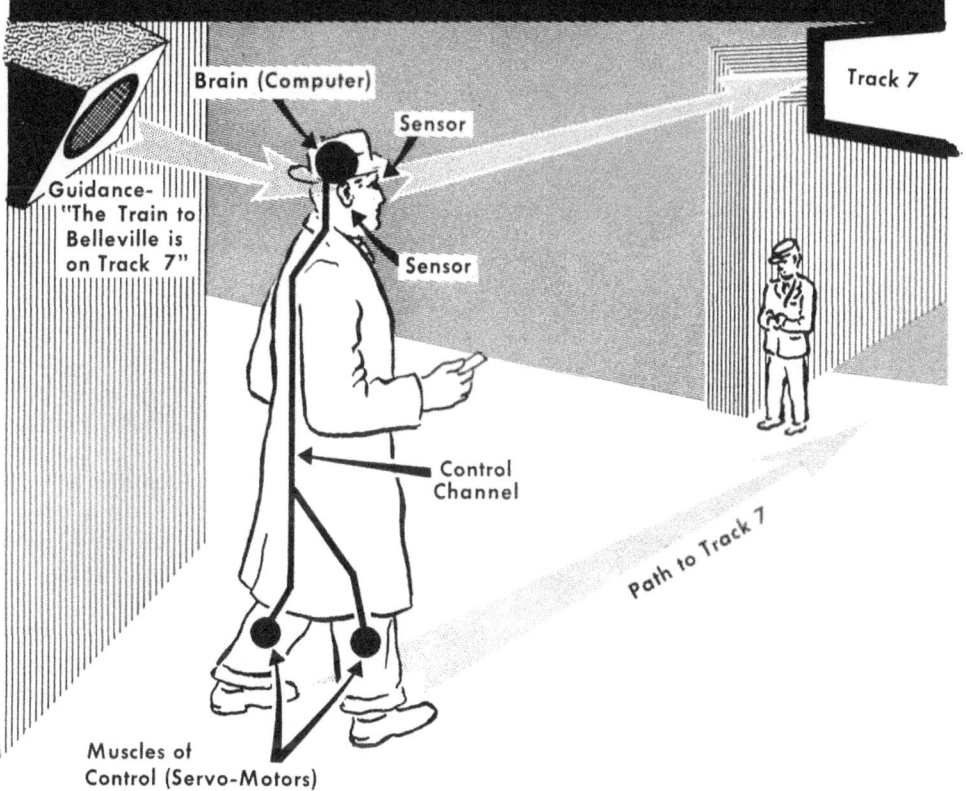

Brain (Computer)

Sensor

Track 7

Guidance-
"The Train to
Belleville is
on Track 7"

Sensor

Control
Channel

Path to Track 7

Muscles of
Control (Servo-Motors)

Earlier we showed the analogy of the human system to a control-loop servo system. Now we can add to this example the concept of guidance to further illustrate how a control system can have a guidance loop tied to it. As we all know, man can be directed through the sensing elements of sight and hearing to follow directions. The feedback to make any necessary corrections of path could come through either of these sensors.

The human analogy can be extended to show how the computer or *brain* would respond to internal signals indicating a need to take certain corrective action through the muscles as a servo-control channel. The nerve system would provide the feedback essential to the smooth execution of the necessary action.

The Elements of Guidance (contd.)

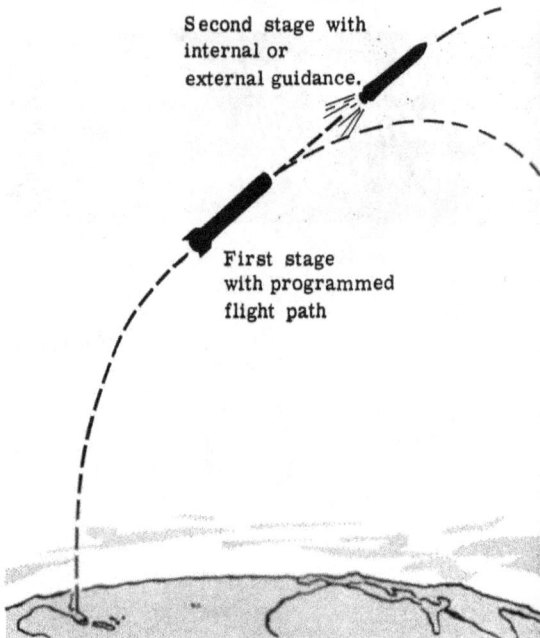

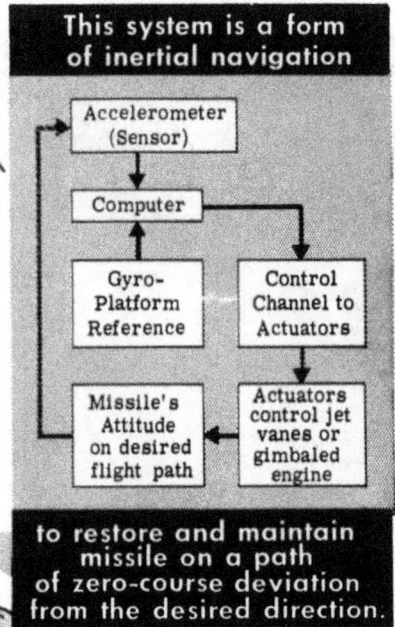

Second stage with internal or external guidance.

First stage with programmed flight path

This system is a form of inertial navigation

Accelerometer (Sensor)

Computer

Gyro-Platform Reference

Control Channel to Actuators

Missile's Attitude on desired flight path

Actuators control jet vanes or gimbaled engine

to restore and maintain missile on a path of zero-course deviation from the desired direction.

Just as the human brain is essential to the determination of the actions to be taken, the computer in the missile system is essential to compare the actual course to a reference, and calculate the inputs to the servo-control elements for guidance over a desired course. Suppose that we programmed the first stage of a two-stage rocket into a curved trajectory as indicated above. When the first stage burns out and falls away, the second stage will be aimed in a certain direction. We may wish to guide this portion of the rocket into some desired trajectory, either as a ballistic missile or to a point where it could be taken over by some form of missile guidance. To do this we could equip the second stage with a gyro-stabilized platform onto which is mounted one or more accelerometers which sense any motion away from the direction into which this stage must move.

The combination of the gyroscopes in a stable platform and the accelerometers form closed-loop corrective networks to detect any deviation of the missile from a desired direction, and together with a computer-amplifier and control-actuators tend to restore the missile to a path of zero-course deviation. This arrangement is a form of inertial navigation which will be examined more thoroughly in later portions of the book.

The insertion of radar command or beam-rider data into the computer can also be used to supplement the inertial data. In the case of short-range defensive missiles, much of the guidance is from radar sources, and the inertial aspect is less significant.

RADIO AND RADAR COMMAND LINKS

Introduction to Radio and Radar Command Links

In this era of missile and space-vehicle control and guidance, radio and radar command links play an essential role in a variety of applications. As progress is made toward the conquest of space, uses will range from model airplane control near the surface of the earth out to the exchange of inter-orbital commands between space vehicles.

Fundamentally, the basic situation is similar throughout the wide range of command data transfers. Signals are transmitted from one point in space to a moving craft to direct certain actions which will influence its course and/or rate of travel. Of course, the simple data link required to produce single-direction rudder control for a model aircraft is very elementary when contrasted with the data-complex necessary to simultaneously direct several air-defense missiles into the correct courses. Also, the systems necessary for the transmission of a wide variety of data from a missile or space vehicle back to its base often tax the limits of information-handling capacity in a given frequency band to the utmost.

However, this phase of communications between a space vehicle and its base — with the primary transmission function in the space vehicle or missile — is generally classed as *telemetering*, and will be dealt with later. Now, we shall confine our considerations to the transfer of command data from a control point to one or more remote points. This is done with a primary transmission facility at the control point to direct actions in the craft at these remote points.

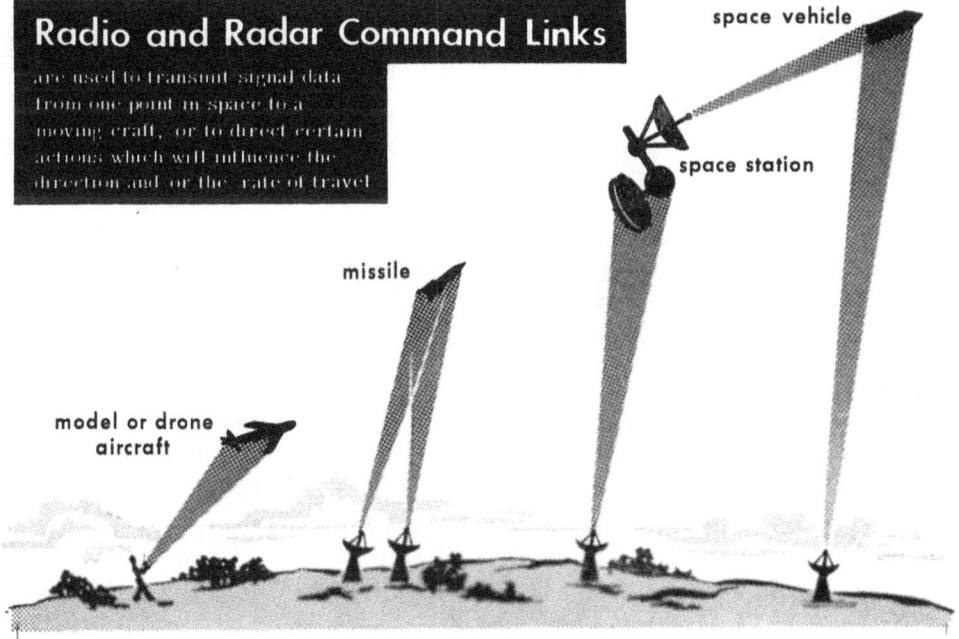

Radio and Radar Command Links are used to transmit signal data from one point in space to a moving craft, or to direct certain actions which will influence the direction and/or the rate of travel

space vehicle

space station

missile

model or drone aircraft

Radio Control of Model Craft

Model Airplane Transmitter and Receiver

A single tube, crystal-controlled transmitter provides highest possible output.

Simple, reliable hard tube receiver controls model as far as you can see it.

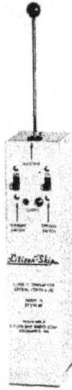

During 1958 five additional frequencies just below 27.255 mc were assigned by the FCC for model control purposes

Techniques for the control of model aircraft by radio command links have been developed over a period of more than 25 years. In fact, radio-controlled flights were investigated as far back as World War I. Prior to World War II a substantial effort was put forth by licensed amateurs in their uhf activity and the control of pilotless aircraft became a subject for development in many countries. Through the years single-tube transmitters — manually keyed for single-channel rudder control — and receivers of the superregenerative type, featuring either a single *hard* or *gas* tube, have enjoyed a high degree of popularity among both novices and experienced operators. The more technically inclined amateurs have gone in for more elaborate multi-tube pulsed systems, but the single-tube circuits are still popular today. After all, they provide an easy approach to model flight control, and permit mastery of the basic principles before refinements are pursued.

The simple circuitry shown provides for the control of only one element of a model aircraft, such as the rudder, which is positioned from right to left by varying the length of the transmitted pulses manually.

Radio Control of Model Craft (contd.)

Other elements may be controlled by additional channels conveying the same general type of pulse signals. Operation by the average nonmilitary model-flyer is conducted in the Citizens bands of 27.255 mc and 465 mc. Such operation eliminates the necessity for a regular amateur operator's license, requiring only the completion and filing of a form (generally enclosed in the control transmitter package) to acquire a Citizens station license. The 27.255-mc band is actually a spot frequency with a permissible deviation of 10 kc. Although crystal control is required, the operator may construct his own transmitter of 5 watts or less output at this frequency. At 465 mc the transmitter must be an approved factory built unit. Great improvement in the compactness and the power consumption of the airborne receiver can be realized by the use of transistors.

MULTI-SIMULTANEOUS 8-CHANNEL RECEIVER

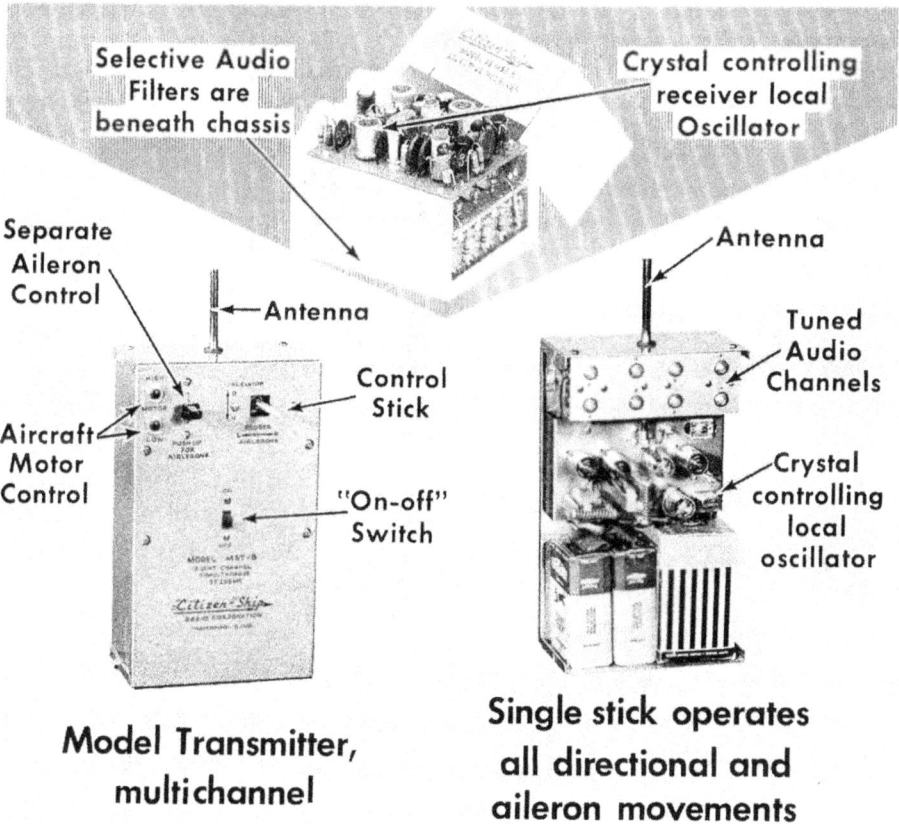

Selective Audio Filters are beneath chassis

Crystal controlling receiver local Oscillator

Separate Aileron Control

Antenna

Control Stick

Aircraft Motor Control

"On-off" Switch

Antenna

Tuned Audio Channels

Crystal controlling local oscillator

Model Transmitter, multichannel

Single stick operates all directional and aileron movements

Radio Control of Model Craft (contd.)

To simulate the stick type control of the piloted plane, and to obtain proportional control relative to the movement of such a simulated *stick*, a so-called *beep box* is often employed. The position of the control stick or knob can be arranged so that it controls the pulse output of a motor-driven controller. Then, by moving the stick to the *right* or *left*, the pulse durations can be controlled so that the aircraft rudder or other surface element is made to move a proportional amount.

During World War II, the Germans developed a rather unique pulse controller which they called a *knupple*. By using a drum which is half-conducting and half-insulated, and shifting the control stick relative to these portions of the drum, pulses of varying lengths can be sent to give *right* and *left* rudder as well as *neutral* positions. The drum is motor-driven at 200 to 300 rpm to give a continuing series of pulses of the order of 5 cycles or a steady signal depending upon the position of the control stick.

The great national efforts in Germany as well as in other countries during World War II led to the development of more elaborate command-signal techniques. One which saw considerable use is the multichannel tone method. In this arrangement the r-f carrier was modulated by one or more audio signals. Although variations in the amplitude of the audio signals could serve to provide a basis of control, it was found preferable to pulse the audio tone in much the same fashion as the (r-f) carrier was pulsed in the previous examples mentioned above. The pulsed signals are decoded at the receiver to provide output suitable for a servosystem in the aircraft, glide bomb, or missile to regulate the rudder or other flight control elements.

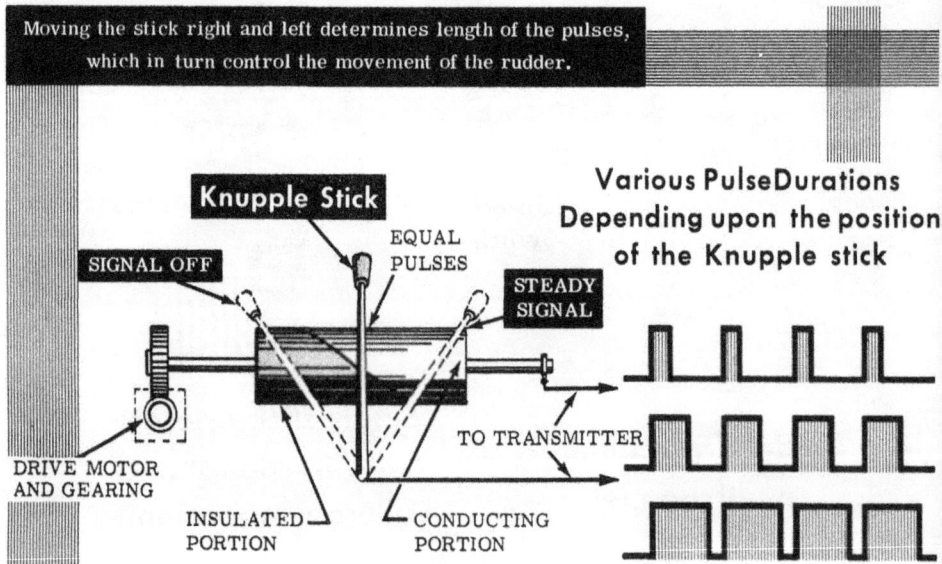

Moving the stick right and left determines length of the pulses, which in turn control the movement of the rudder.

Knupple Stick

SIGNAL OFF

EQUAL PULSES

STEADY SIGNAL

Various Pulse Durations Depending upon the position of the Knupple stick

TO TRANSMITTER

DRIVE MOTOR AND GEARING

INSULATED PORTION

CONDUCTING PORTION

Radio Control of Model Craft (contd.)

As was mentioned above, several separate r-f channels may be utilized to control a variety of functions. If the number of functions to be controlled is larger than can be conveniently handled in the r-f spectrum available for separate channels, audio tones may be modulated onto each r-f carrier to expand the number of control channels considerably.

For instance, one German system — known as the *Strasburg E230* — used 18 r-f channels with locking selectors in the frequency range of 48 to 50 mc with a channel spacing of 100 kc. It also used amplitude-tone modulation of 1, 1.5, 8, and 12 kc. With such an arrangment, several air-launched bombs could be controlled simultaneously. One frequency band assigned for present-day drone operation is 406 to 550 mc.

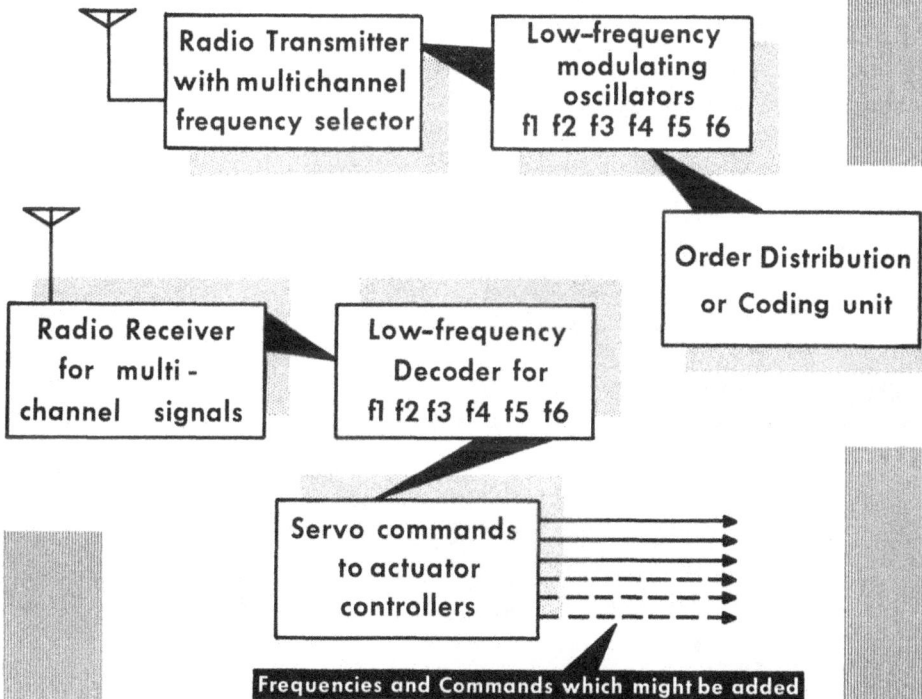

The "Kehl-Strassburg" system controlled glide bombs and target drones of World War II

The Transition to Weapon Control

Up to this point we have dealt with types of radio command signalling which would seem relatively simple to the present-day missile-guidance designer. Yet, we have covered some basic points which serve to illustrate certain common aspects of all types of command-guidance. In dealing with some of the basic elements of model-aircraft control, it is our intent to present a background for the understanding of the general underlying philosophy of radio and radar command link techniques.

Practically all of the arrangements we have described use pulse modulation — whether they be in the r-f or a-f range. As we examine the techniques for radar command guidance, we shall see that the pulse philosophy still dominates the scene even though it may be of microsecond duration rather than at low a-f rates. Quite effective pilotless aircraft and glide-bomb control and guidance systems have been constructed by utilizing long-duration pulses of r-f or pulsed audio tones. The possibility of placing large amounts of information in relatively narrow frequency bands through the use of microsecond pulse techniques has become a principle factor in most present-day command-guidance systems.

Before we examine the manner in which the pulse techniques are utilized in conjunction with radar signals, we should study some pulse-modulation theory and some of the related basic circuits. Even before studying the general theory of pulse-modulation techniques, we shall compare them with multichannel subcarrier operation, and show the advantages of such systems.

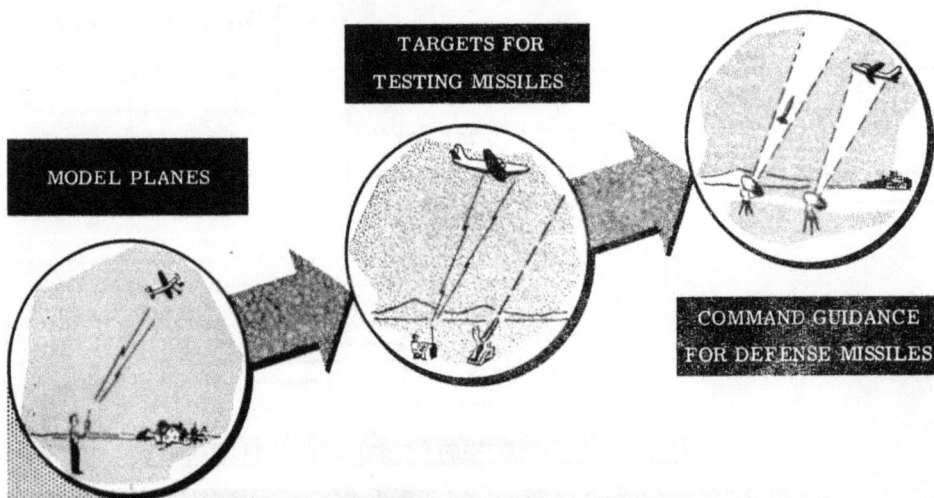

TARGETS FOR TESTING MISSILES

MODEL PLANES

COMMAND GUIDANCE FOR DEFENSE MISSILES

The Transition to Weapon Control by Command Guidance

Frequency-Division *vs* Time-Division

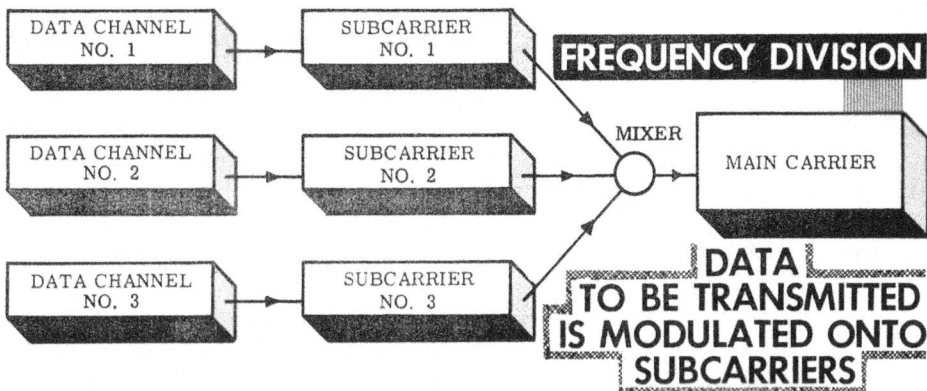

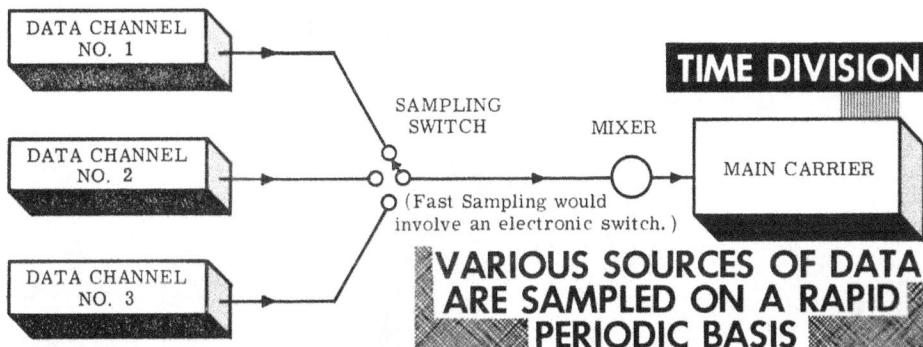

FREQUENCY DIVISION

DATA TO BE TRANSMITTED IS MODULATED ONTO SUBCARRIERS

TIME DIVISION

VARIOUS SOURCES OF DATA ARE SAMPLED ON A RAPID PERIODIC BASIS

Two broad courses are open to the solution of an increased data-handling requirement in a command-guidance link. The data to be transmitted can be modulated onto subcarriers either in sinusoidal or pulse form. This method is known as the *frequency-sharing technique* or *frequency-division.* Within a given bandwidth available for transmission of a data complex, the frequency spectrum can be shared by dividing it into a number of modulated subcarriers. At the receiver the subcarriers are separated by a detection system. The modulation is removed from each of them to provide the signals for the desired control actions.

The second course is to sample the various sources of data at the transmitter on a rapid periodic basis. The pulses representing data from several sources are spaced along the time axis in such a manner that they do not overlap or interfere with each other. These separate pulse-trains are modulated onto the r-f carrier at the transmitter, and are separated at the receiver by *gating* circuits which bring out each series of pulse-data into separate channels to provide signals for the desired control actions. This system is also referred to as *pulse-time multiplexing* or *time-division.*

Pulse Modulation

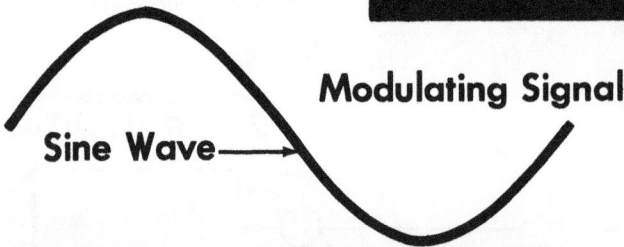

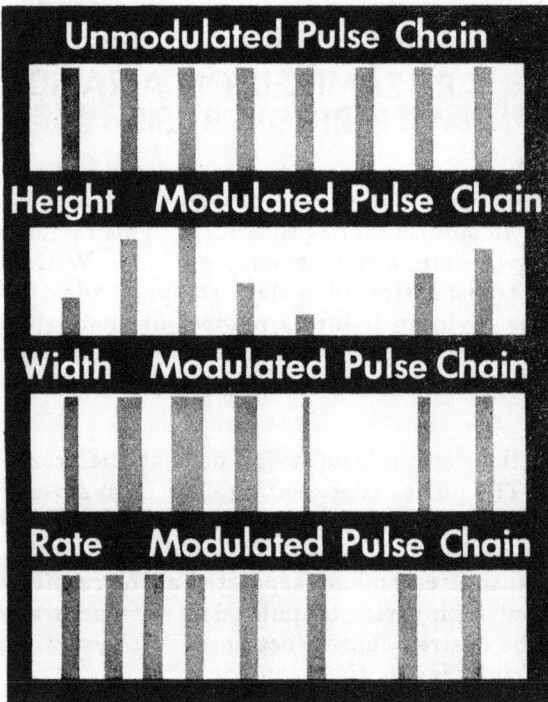

There are as many pulse modulation systems as there are variable parameters in a train of pulse signals. For instance, a pulse may be varied in amplitude, duration, and shape. Additional parameters may be established by the spacing between pulses, the number of pulses of various widths, and by the presence and omission of pulses in a pulse train. The more common means for modulating pulses are designated as amplitude, width, time, and frequency modulation. In considering the methods of achieving these forms of modulation as well as their decoding at the receiver, we shall describe a number of circuits which are widely used in the application of pulse techniques.

PULSE MODULATION

Modulating Signal

Sine Wave ⟶

Unmodulated Pulse Chain

Height Modulated Pulse Chain

Width Modulated Pulse Chain

Rate Modulated Pulse Chain

In all these pulse modulating methods a particular property of a train of pulses is caused to change with the variation of the quantities to be represented.

Pulse Amplitude Modulation (PAM)

Pulse Amplitude Modulation (PAM) involves the variation of pulse height in relation to a modulating signal. It may take one of two forms — either, unidirectional or bidirectional operation.

UNIDIRECTIONAL PAM

BIDIRECTIONAL PAM

Although amplitude pulse modulation is rather easily realized by simple circuitry in which the pulses are fed to one grid and the modulating signals are fed to another grid of a multi-element tube, it is not as widely used as it might be for missile command guidance because it has a number of disadvantages. For instance, it requires linear equipment, and does not permit the use of limiting to the fullest extent required for noise reduction.

Simple Example of Unidirectional PAM Circuit

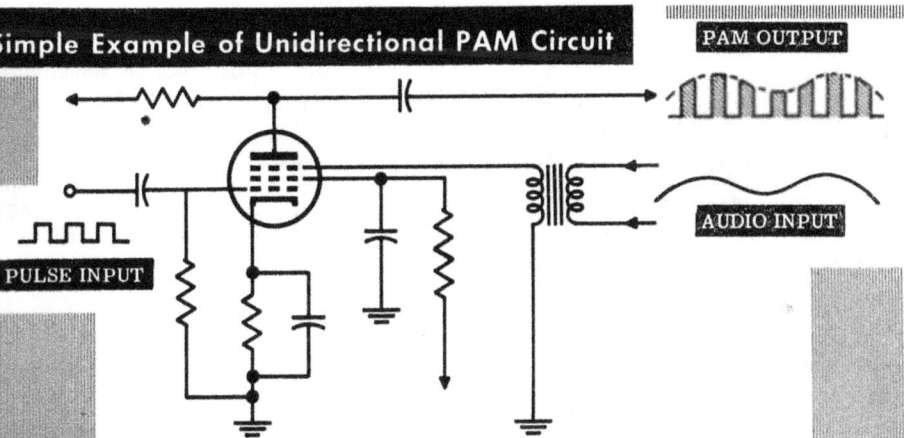

PAM OUTPUT

AUDIO INPUT

PULSE INPUT

Pulse amplitude in the plate circuit is controlled by suppressor grid modulation.

Pulse Width Modulation (PWM)

Pulse Width Modulation (PWM) is produced by varying the width of a series of pulses along the time axis. A technique for realizing this method involves the simultaneous application of sawtooth and sinusoidal waveforms to the control grid of a modulator tube.

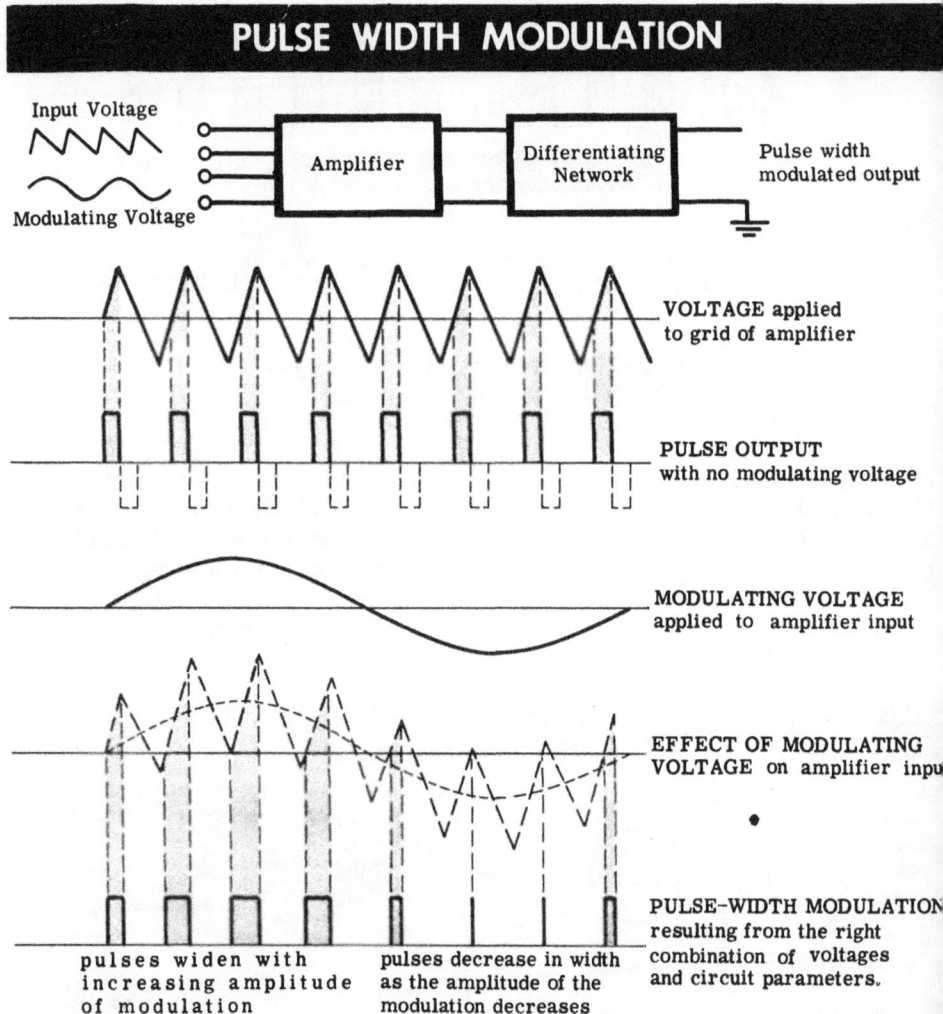

PULSE WIDTH MODULATION

Input Voltage

Amplifier

Differentiating Network

Pulse width modulated output

Modulating Voltage

VOLTAGE applied to grid of amplifier

PULSE OUTPUT with no modulating voltage

MODULATING VOLTAGE applied to amplifier input

EFFECT OF MODULATING VOLTAGE on amplifier input

PULSE-WIDTH MODULATION resulting from the right combination of voltages and circuit parameters.

pulses widen with increasing amplitude of modulation

pulses decrease in width as the amplitude of the modulation decreases

Although in pulse width modulation the variation in the average power in the carrier is approximately the same as in pulse amplitude modulation, the pulse amplitude remains constant facilitating the use of limiting and clipping to eliminate noise peaks.

RADIO AND RADAR COMMAND LINKS

Pulse Time or Pulse Displacement Modulation (PTM or PDM)

Pulse Time or Pulse Displacement Modulation (PTM or PDM) involves
the timing or displacement of pulses of a fixed width and amplitude in accordance with a modulating signal. It is closely related to pulse width
modulation in that it may be obtained by simply differentiating the pulse
width modulation signal.

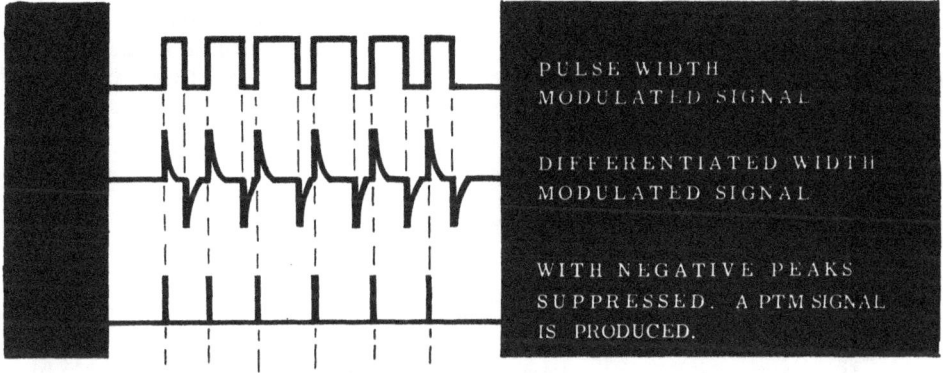

PULSE WIDTH
MODULATED SIGNAL

DIFFERENTIATED WIDTH
MODULATED SIGNAL

WITH NEGATIVE PEAKS
SUPPRESSED. A PTM SIGNAL
IS PRODUCED.

It should be noted that differentiation of a rising or decaying pulse may be
accomplished by a simple RC network.

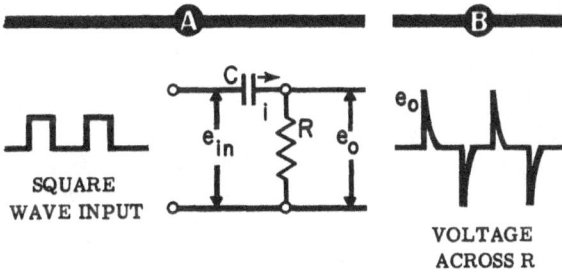

SQUARE
WAVE INPUT

VOLTAGE
ACROSS R

Voltage across R is the rate
of change or differentiation of
the input across C and R,
while C will accumulate
charges from the pulses and
give their integration over a
period of time.

The voltage across R rises promptly during the rising wavefront of the
square wave pulse (A), and it decays rapidly as the capacitor C charges to
give the waveform shown at (B). The time constant of the R-C network
determines the exact waveshape of the output signal at B. It is a differentiator because its output represents the rate of change of the input voltage.
Rate of change of any function represents its differentiation mathematically.
If the time constant of the circuit is very small so that the flow of current
is controlled almost entirely by the capacitance, $i = Cde_{in}/dt$ and the output
voltage across R is expressed by $e_o = RC\, de_{in}/dt$.

Pulse Frequency Modulation and Pulse Count Modulation

Before we show how time-division multiplexing of several separate commands and messages is accomplished, we shall describe briefly two other forms of pulse modulation — pulse frequency modulation (PFM) and pulse count or code modulation (PCM). In PFM, the frequency of the pulses varies in accordance with the modulating signal. In PCM the groups of pulses are varied in number in accordance with the input signal level. Typical operation involves the use of binary code, which is explained later in the book. In many applications, the input signal will be available in amplitude-modulation form. In a coder or converter, corresponding pulses spaced in accordance with binary code for each amplitude level are produced. The

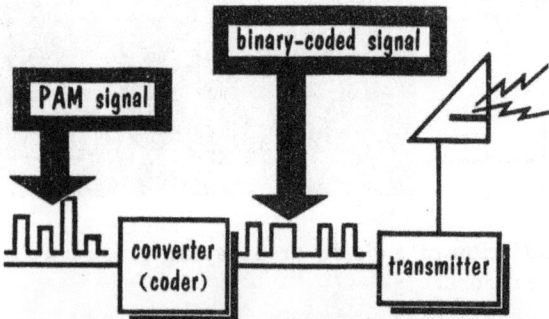

A BASIC PCM SYSTEM

Binary Output To The Transmitter May Be

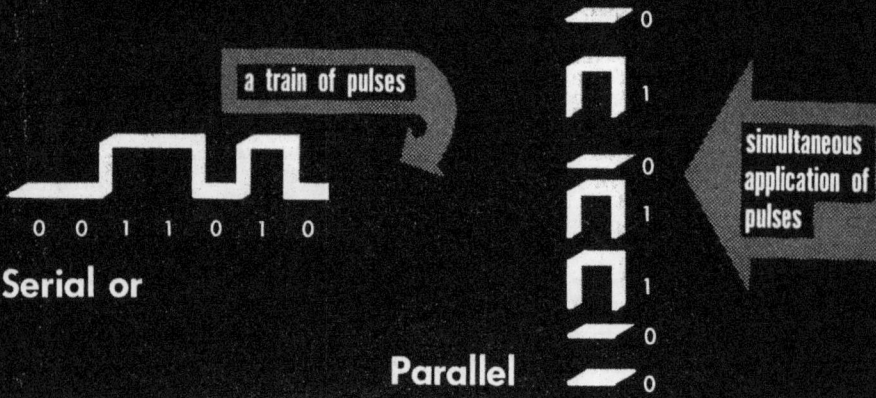

a train of pulses

simultaneous application of pulses

0 0 1 1 0 1 0

Serial or

Parallel

output of the converter may be in either serial or parallel form. In serial form, the bits are in a continuous train, while in parallel form the bits are available simultaneously.

Pulse Time Multiplexing

As was mentioned previously the use of pulse modulation techniques permits the sampling and transmission of several commands or signals over one carrier in contrast with the use of multiple channels or subcarriers in so-called frequency-sharing systems. Although pulse time modulation undoubtedly provides a maximum amount of space between pulse trains to intersperse additional messages, we shall illustrate the possibilities with pulse amplitude modulation for the sake of simplicity.

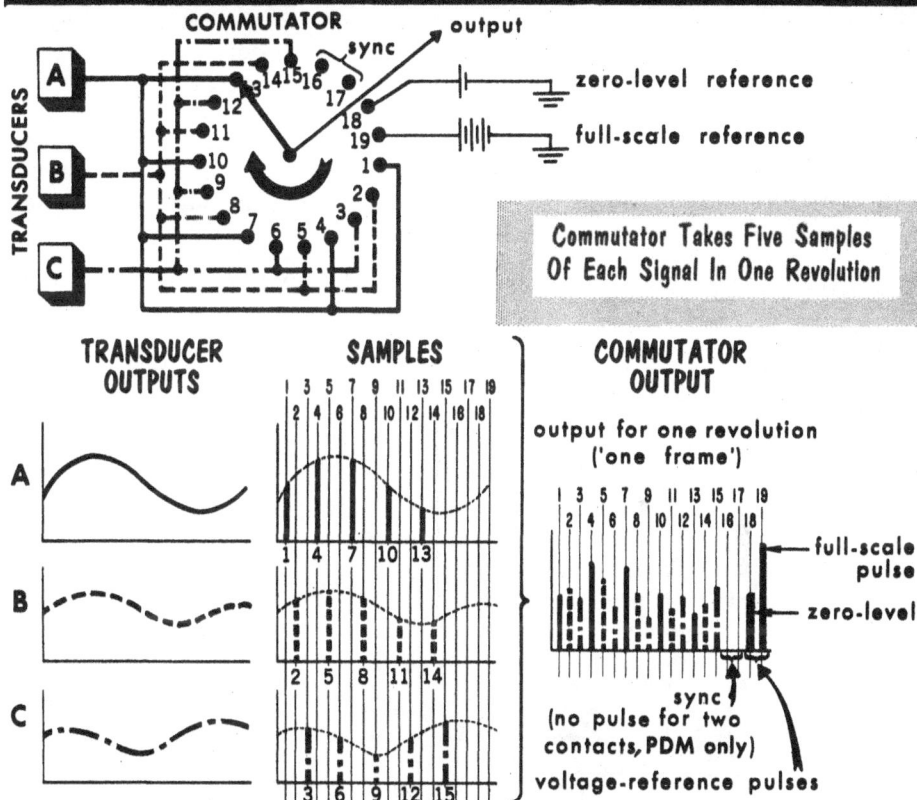

TIME-DIVISION MULTIPLEXING OF THREE SIGNALS

Commutator Takes Five Samples Of Each Signal In One Revolution

If a message is sent by the amplitude modulation of pulses a, b, c, and d as shown in the above figure, it is quite apparent that with one carrier frequency we can intersperse other messages at positions 1 to 3. By utilizing a synchronizing pulse, which would be of greater amplitude and unvarying with modulation signal, it is possible to keep pulse trains separated and to detect the message from each on a time-sharing basis.

Pulse Time Multiplexing (contd.)

The voltages of each pulse train to be applied to a transmitter stage for modulation are fed through channel-gate tubes. A typical example is one in which the screen-grid voltage of the gating tubes is pulsed in sequence while the modulation signals are applied to the separate control grids. The proper time position of each pulse may be established by a tapped delay line or by using channel gate oscillators. These are first triggered by the synchronizing pulse and sequenced at proper intervals. Such channel gate oscillators are usually a form of multivibrator circuit.

At the receiver as well as at the transmitter, gating circuits must be employed to separate pulse-train messages. The transmission and separation of the synchronizing signal is very similar in principle to that employed in television transmission to establish correct timing for horizontal and vertical sweep voltages. In the case of pulse width modulation, pulse time modulation, and other forms where the amplitude of the pulse is not varied, limiters are employed at the receiver ahead of any pulse separation functions. However, beyond that point in the circuitry, sync signal separation and channel gating must be employed to separate the messages on the adjacent pulse trains.

An important consideration in any pulse modulation system is the rate at which intelligence can be transmitted. If the given bandwidth is ΔF and the signal-to-noise ratio P/N, where P is the signal power and N is the noise power, this rate is expressed by the formula $C = (\Delta F) \log_2 (P + N)/N$ bits/second. During time, T, the total intelligence which can be transmitted is $CT = (\Delta F) T \log_2 (P + N)/N$ bits.

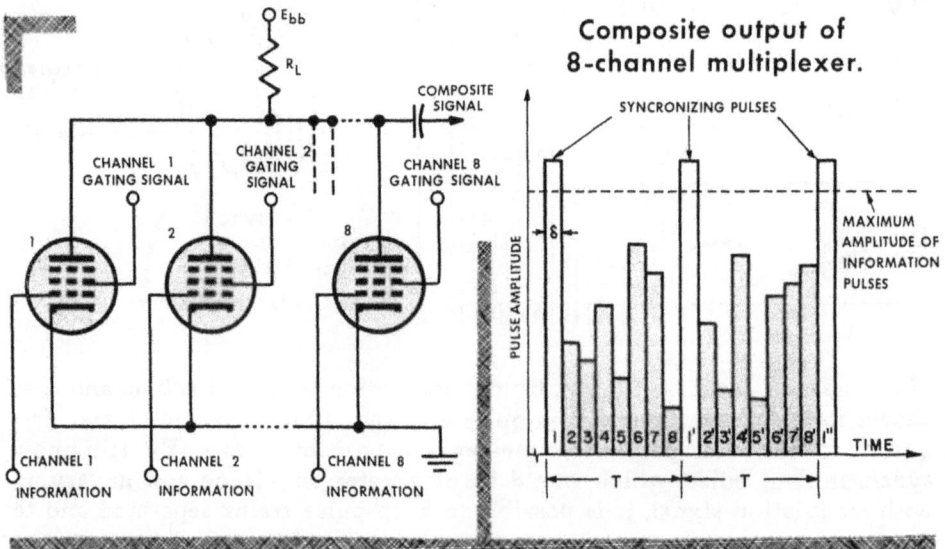

Composite output of 8-channel multiplexer.

Radar Command Guidance .

Radio Command Guidance

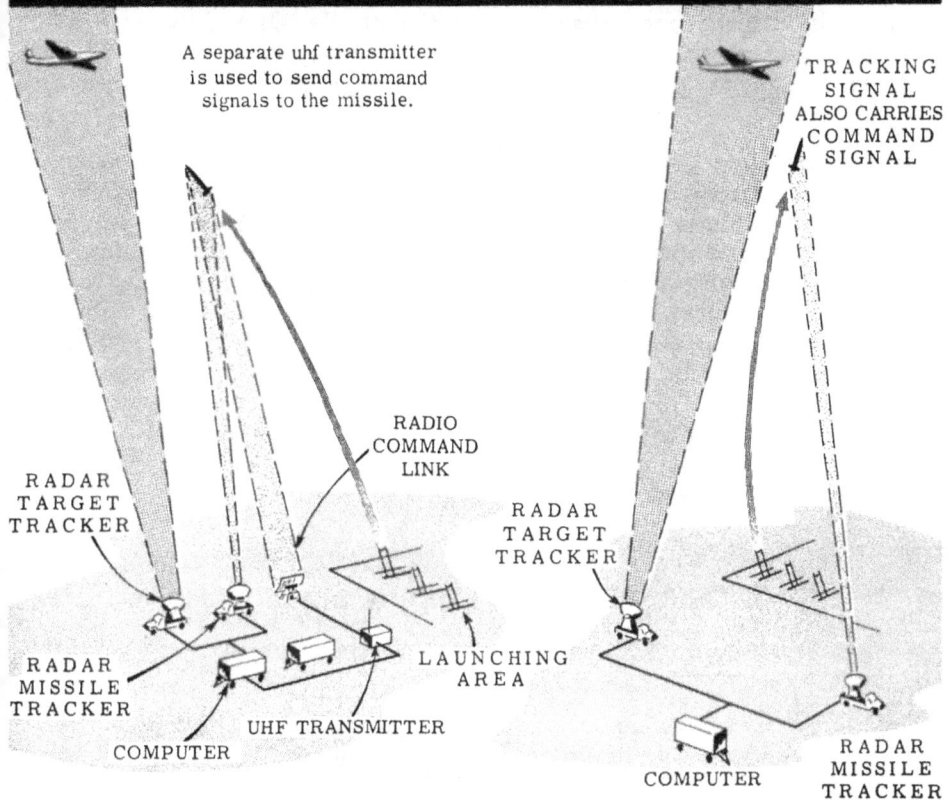

A separate uhf transmitter
is used to send command
signals to the missile.

TRACKING
SIGNAL
ALSO CARRIES
COMMAND
SIGNAL

RADIO
COMMAND
LINK

RADAR
TARGET
TRACKER

RADAR
TARGET
TRACKER

RADAR
MISSILE
TRACKER

LAUNCHING
AREA

UHF TRANSMITTER

COMPUTER

COMPUTER

RADAR
MISSILE
TRACKER

Now that we have expressed some of the fundamental concepts of the utilization of pulse techniques for command signalling, we shall look at the manner in which these methods may be applied to the radar equipment in a command-guidance system. In the earlier forms of this type of guidance radar tracked the target and the missile, and fed data to a computer to determine the correct commands to be forwarded to the missile. These commands were then sent over a radio link separate from the radar systems, but of course connected to the computer. Thus, both a tracking radar[1] and a *uhf* transmitter maintained contact with the missile during its ascent to the moving target.

---

[1] A form of radar, known as Monopulse, which uses a phase comparison method — has become popular because of its electronic scanning capabilities.

Radar Command Guidance (contd.)

As command guidance saw further development, it became apparent that radar could be utilized not only for tracking but also for conveying the guidance intelligence from the ground to the missile. One of the parameters of a radar system which can be modulated to convey command intelligence while tracking is the pulse train of radiated energy normally associated with ranging.

The simplest method would be to amplitude-modulate the pulses. Another pulse modulation method to convey command signals via the radar beam involves the variation of the pulse repetition rate (PRR). The necessary variable error-correction information to establish the desired course along one axis may be transmitted by varying the pulse repetition rate by up to 10% of the average frequency. Command signals might be sent also by shifting the PRR to another average frequency. Pulse time modulation (PTM) or pulse position modulation (PPM) may also be employed by transmitting two separate pulses from the radar equipment. One of these pulses is used as a reference, while the other may vary along the time axis, as mentioned in the earlier description of PTM, in accordance with modulation suitable for proportional control.

METHODS OF MODULATING A PULSE TRAIN

Amplitude Variation of Radar Pulses

Variation of Radar Pulse Repetition Rate

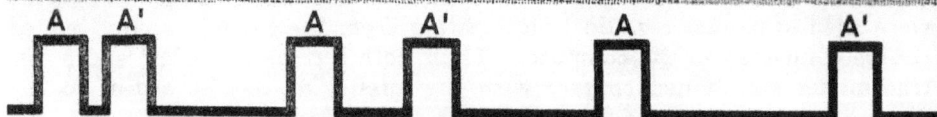

PULSE POSITION MODULATION The Spacing or Position of additional pulse A' relative to Radar Pulse A

Radar Command Guidance (contd.)

In a typical radar system, employing pulse modulation of the radar signal, a beacon is carried in the missile to enhance the return signal over that which could be realized with so-called *skin-tracking* by direct reflections from the missile's surface. The missile beacon responds only to a certain pulse combination from the ground radar and sends back a signal, which is tracked by the ground radar.

In addition to the tracking pulses the radar system transmits two additional pulses A and B. The latter pulse is shifted back and forth in position relative to pulse A in accordance with the amplitude of the command signal so that pulse position modulation is realized. When the pulse modulated signal is accepted at the beacon receiver it is used to trigger a multivibrator. The fixed pulse starts the multivibrator and the variable-position pulse stops it. Thus, pulses of varying width are generated. These are fed through an integrating circuit so that the square wave produces a signal with a sawtooth top. The amplitude variation of this signal can be filtered so that a sinusoidal voltage is developed for a control function of the missile.

Using the Radar Signal for Command Communication

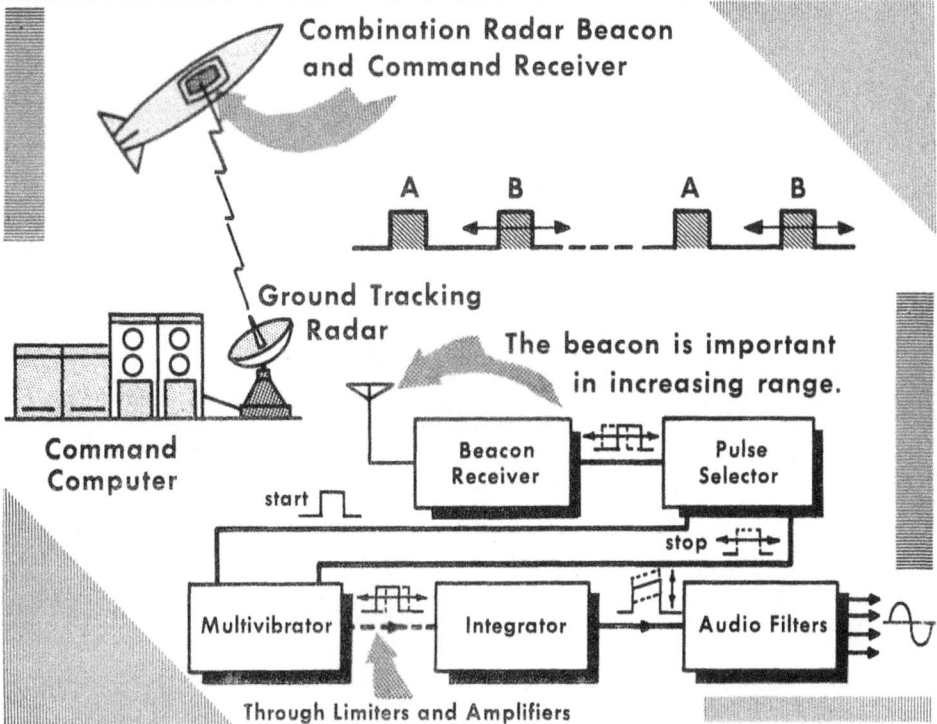

Combination Radar Beacon
and Command Receiver

A   B   A   B

Ground Tracking
Radar

The beacon is important
in increasing range.

Command
Computer

Beacon
Receiver

Pulse
Selector

start

stop

Multivibrator

Integrator

Audio Filters

Through Limiters and Amplifiers

Modern Radar Systems

A typical present-day tracking radar is the AN/FPS-16, which was developed for guided missile test ranges. It is a monopulse type with a somewhat better accuracy than the conical scan radars previously employed.

A two-to-one improvement in range is also provided. The antenna system for this equipment consists of a four-horn monopulse feed. The AN/FPS-16 may be used as a path for two-way communications as well as for location purposes. In such an operation the radar transmitter interrogates a beacon in the missile, whose first response pulse is used to establish range and angle track by the radar. This signal keeps the radar antenna on the target and provides for a continuous position plot on the ground. The drone or missile may receive command guidance through additional pulse signals preceding the ranging pulse. They are coded to match the preset code in the beacon and pulse-position modulated to provide several channels of about 100 cycles bandwidth. As many as 10 or 20 separate pulses can be used, and the bandwidth can be utilized by frequency- or time-division to control several independent functions.

The following comparisons can be made between the AN/FPS-16 and the SCR-584 radar of World War II fame:

Characteristics	SCR-584	AN/FPS-16
Operating Frequency Band	S	C
Peak Power (kw)	250	1000
Antenna Size (diameter, feet)	6	12
Beamwidth (degrees)	4	1.2
Accuracy: Range (yards)	20	5
Angle (mils)	2	0.1

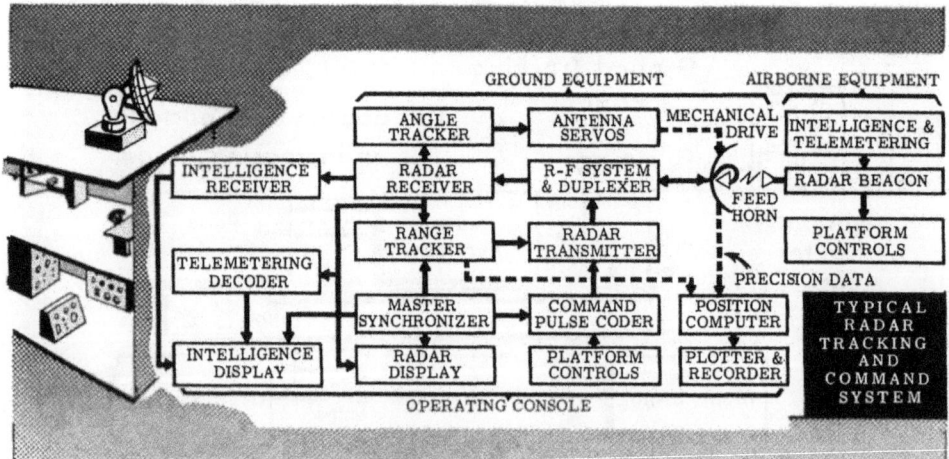

Introduction

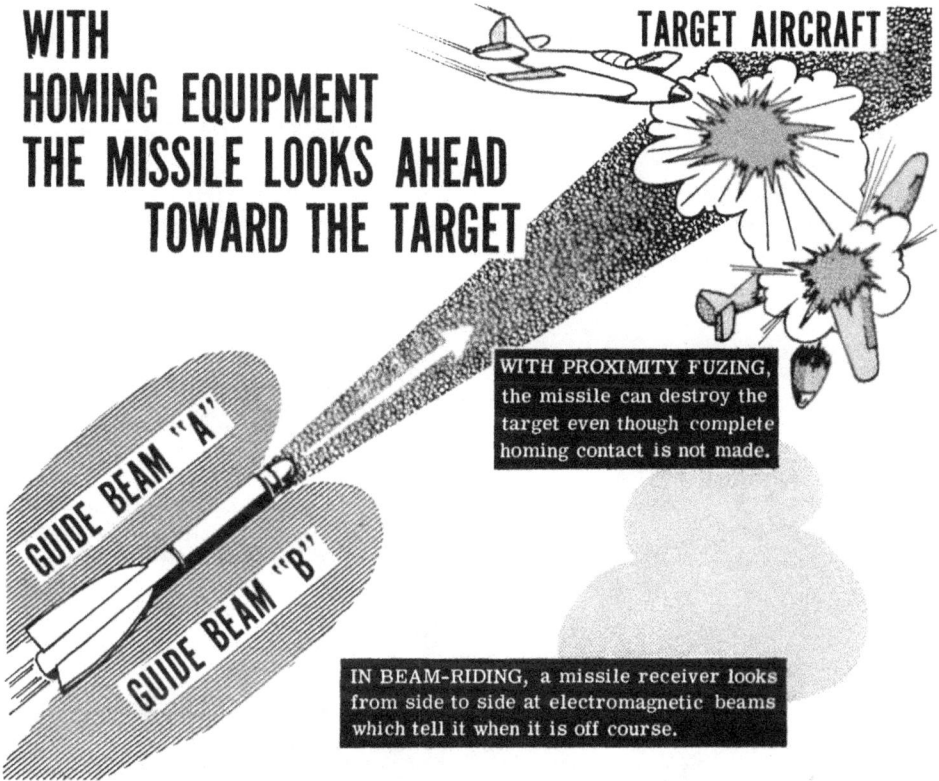

**WITH
HOMING EQUIPMENT
THE MISSILE LOOKS AHEAD
TOWARD THE TARGET**

TARGET AIRCRAFT

WITH PROXIMITY FUZING, the missile can destroy the target even though complete homing contact is not made.

GUIDE BEAM "A"

GUIDE BEAM "B"

IN BEAM-RIDING, a missile receiver looks from side to side at electromagnetic beams which tell it when it is off course.

Previously we dealt with radio and radar methods of providing guidance from sensing devices external to the missile. We shall now deal primarily with radio, radar, and infrared techniques for providing guidance from sensing arrangements within the missile. Later we shall consider internal techniques which employ celestial and inertial sensors. Guide beam, homing, and proximity applications were developed to some degree during World War II. Since that time they have been improved extensively for defensive missiles operating against moving targets — particularly high-speed aircraft.

In the future, further developments especially in homing techniques are indicated in the solution of antimissile defense problems as well as space rendezvous between orbiting vehicles. Homing and proximity techniques may be applied independently or in conjunction with other basic forms of guidance, such as command systems. For example either guide-beam or command techniques may bring the missile onto the proper course with the principal application of homing and/or proximity fuzing as the missile comes into close range of the moving target.

Beam-Riding Guidance

Missiles utilizing guide-beam information to establish their correct position relative to a target area are called *beam riders*. They may be launched either from surface bases or an aircraft, and require that a radio system capable of establishing a guide path, or a radar capable of tracking a target, be available to establish a suitable radiated signal-pattern in which the missile can travel toward the target. The missile contains a receiver for detecting deviations from the proper position in the guide-beam pattern, and from this data establishes the proper flight path without any command signals from the guiding source.

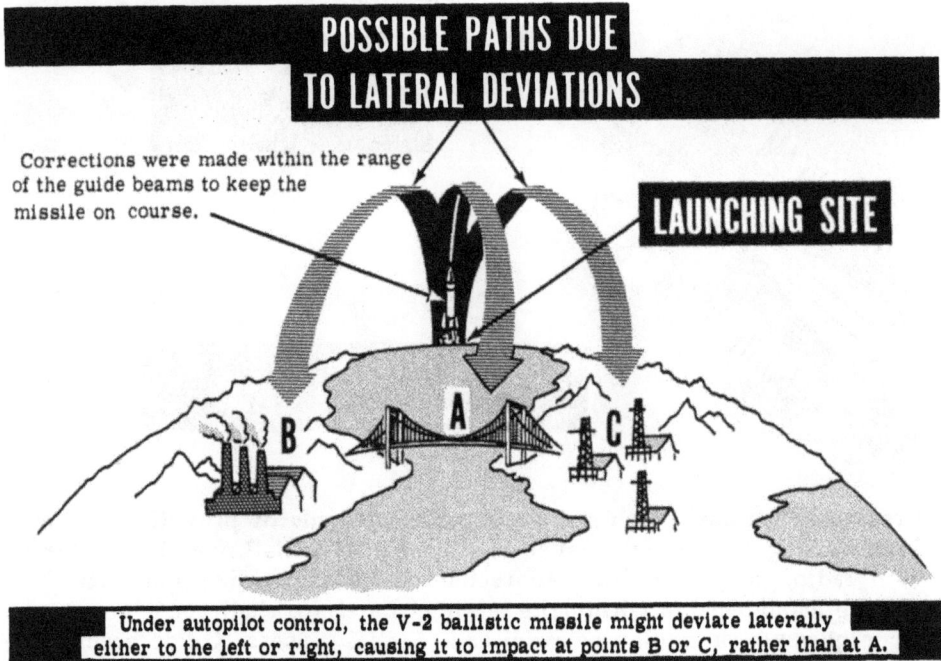

POSSIBLE PATHS DUE TO LATERAL DEVIATIONS

Corrections were made within the range of the guide beams to keep the missile on course.

LAUNCHING SITE

Under autopilot control, the V-2 ballistic missile might deviate laterally either to the left or right, causing it to impact at points B or C, rather than at A.

One of the earliest applications of the guide-beam technique was made during World War II in V-2 guidance utilizing a radio technique, similar to that developed in the mid-1930s for the blind landing of aircraft. Although the Germans often employed a self-contained nonradiating form of inertial guidance for many V-2 launchings, especially when the jamming of radio or radar signals appeared likely, they also developed, and sometimes employed, a radio guide-beam system to correct for lateral deviations from the desired path. With this method they obtained greater accuracy at the target. However, inertial guidance was not fully developed for higher accuracies at that time, so a comparison of guide beam versus inertial guidance would not necessarily be the same today as it was in World War II.

Beam-Riding Guidance (contd.)

In the missile application the blind landing technique was essentially reversed to place an object into a position above the ground rather than on a landing strip. In the V-2 guide-beam system a ground-based transmitter, operating at about 50 mc, was positioned at a distance of about 6 miles behind the launching site. The transmitter fed two horizontal dipoles positioned at a distance of about 600 feet from each other. This feed was alternated between the two antennas at a switching rate of 50 cycles and the beam characteristics *a* and *b*, were obtained alternately.

The signal was also modulated by two additional audio signals — 5000 cycles in beam *a* and 7000 cycles in beam *b*. When the missile was to the left of the desired course, the 5000-cycle modulation predominated at the missile receiver. When to the right of the desired course, the 7000-cycle modulation was the stronger. The signal received at the missile was demodulated down to the 50-cycle modulation. The resulting envelope was passed through a filter for that frequency. At the receiver output in the V-2 rocket a voltage was obtained whose positive or negative sign depended upon the lateral position (right or left) of the rocket, and was proportional to its angular position Θ.

This guide-beam system was developed for 10 different radio frequencies to avoid interference from countermeasures.

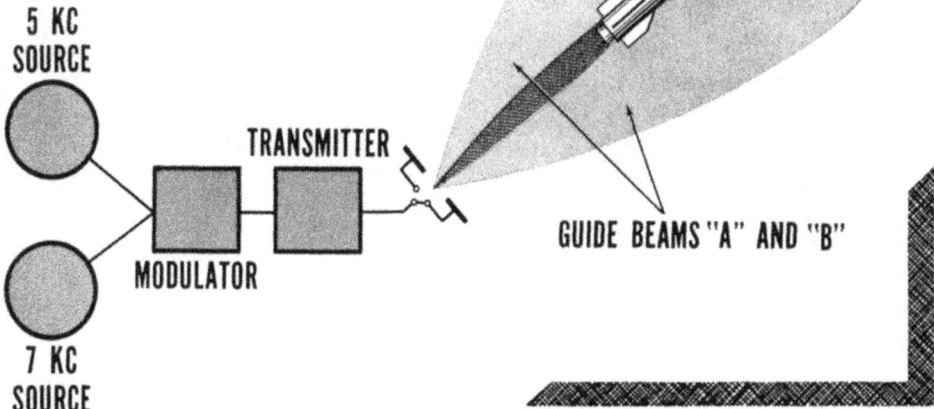

RECEIVER IN MISSILE PROVIDES
CONTROLLING VOLTAGES
WHICH VARY WITH ANGLE θ

5 KC SOURCE

7 KC SOURCE

MODULATOR

TRANSMITTER

GUIDE BEAMS "A" AND "B"

Radar Beam-Rider Guidance

In the application of the beam-rider technique to systems employing radar tracking of a target, the initial guidance phase must depend upon an autopilot in the missile to keep it in a position where it can be captured by the guide beam. However in this technique the guiding radar equipment must provide two essential elements to the missile; (1), the required course to the target, and (2), a reference which enables the missile-circuits to measure the weapon's position in space with respect to this course.

The path along which the missile must fly is defined by the tracking operation of the radar. Prior to launching and throughout the boost phase of the missile, the beam-radar antenna is directed automatically so that the center of its radar-beam pattern points to the target. The second element that enables the missile to determine its position in space relative to the target is provided by guidance signals from the beam received at the missile.

The beam is rotated in a conical pattern in space to give position information to the missile receiver. By demodulating the signal from the radar beam, which is rotated at a low audio rate from 50 to 75 cycles, we may determine the deviation of the missile from the center of the scan pattern by determining the amplitude of the scanning-rate signal. By making a phase-comparison between an audio reference signal in the missile and the scanning signal, it is possible to determine the angular displacement of the missile around the beam-pattern. By applying these deviation signals through the proper control-channels the missile can be brought back to the center of the conical beam and into the correct direction toward the target.

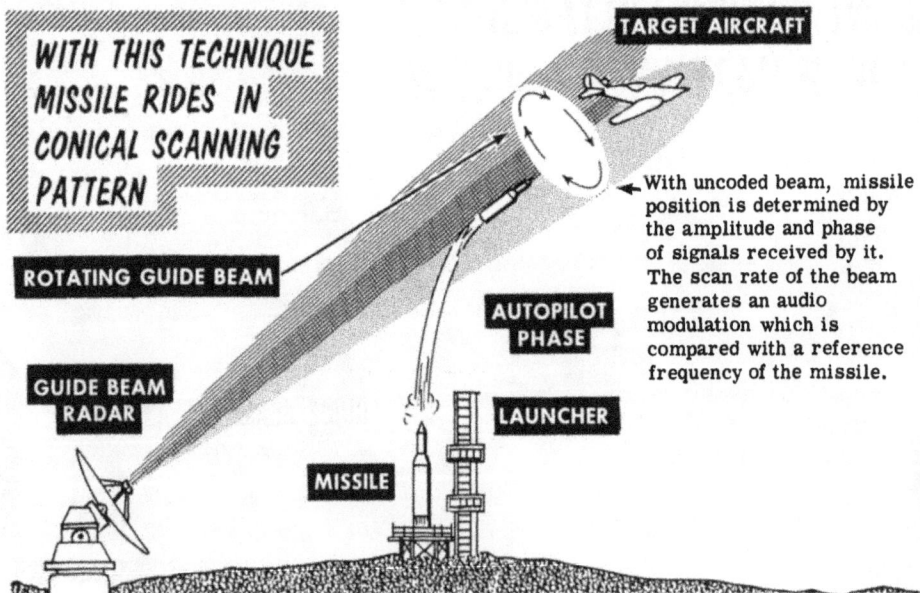

WITH THIS TECHNIQUE MISSILE RIDES IN CONICAL SCANNING PATTERN

TARGET AIRCRAFT

ROTATING GUIDE BEAM

GUIDE BEAM RADAR

AUTOPILOT PHASE

LAUNCHER

MISSILE

With uncoded beam, missile position is determined by the amplitude and phase of signals received by it. The scan rate of the beam generates an audio modulation which is compared with a reference frequency of the missile.

Radar Beam-Rider Guidance (contd.)

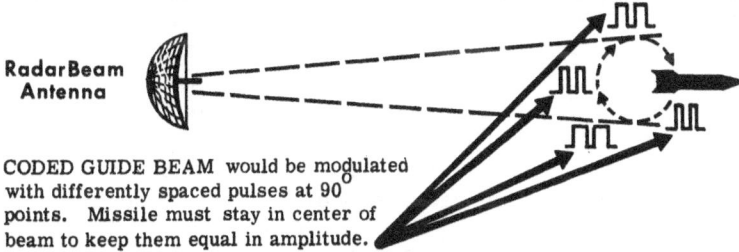

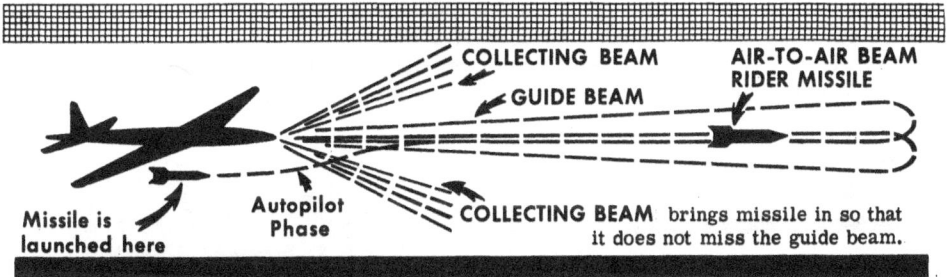

Radar Beam-Rider Guidance

RadarBeam Antenna

CODED GUIDE BEAM would be modulated with differently spaced pulses at 90° points. Missile must stay in center of beam to keep them equal in amplitude.

COLLECTING BEAM

GUIDE BEAM

AIR-TO-AIR BEAM RIDER MISSILE

Missile is launched here

Autopilot Phase

COLLECTING BEAM brings missile in so that it does not miss the guide beam.

Another method of determining the missile's position in the conical scan pattern is to place a pulse code on the radar signal. With this method pulses of varying spacing are placed at 90° intervals around the circle described by the conical scan. By comparing the intensity of these pulses in the missile receiver it is possible to provide control voltages which will send the missile *up* or *down* or to the *right* or *left* to bring it into the center of the conical scan pattern. Although a beam-rider missile depends upon its autopilot to establish its course during the launching phase, it can be brought under control at an early stage in its flight path by providing a *collecting* radar beam in addition to the normal conical-scanning radar beam. This beam subtends a somewhat wider angle, and can be picked up by the missile receiver closer to the launching point than the narrow guide-beam. After receiving the *collecting* signal the missile will intercept the guide-beam as it travels toward the target.

Another refinement of the beam-rider system is sometimes employed. A separate radar equipment tracks the target, and provides follow-up information to the guide-beam radar. This method is more applicable to surface-launched missiles, where excessive maneuvering at low altitudes might be required in a single beam system.

Single radar guide-beam systems have been used for air-to-air missiles. In this case several missiles, fired in sequence, can be guided by one radar. If they are spaced by a slight time interval, no difficulties should be encountered.

Doppler Principles and Applications

For navigation and homing purposes it is often advantageous to employ CW radar. The Doppler shift of frequency offers a medium for extracting information from a CW system. A common effect in nature is demonstrated every day of our lives by the change in pitch of horn signals and whistles from moving vehicles, such as autos and railroad engines. The pitch of the sound is shifted from that present when the vehicle is stationary with respect to the listener. This principle is named for Christian Johann Doppler, who stated in 1842 that "If the distance between an observer and a source of vibration is changing the frequency of vibrations appears to become greater or smaller than the true value — depending on whether the distance is being diminished or increased." By utilizing this principle, the velocity of an object relative to that of a source of vibrations can be determined, if a comparison is made between the frequency of the fixed source and the apparent frequency of the vibrations at the object in motion.

THE DOPPLER EFFECT

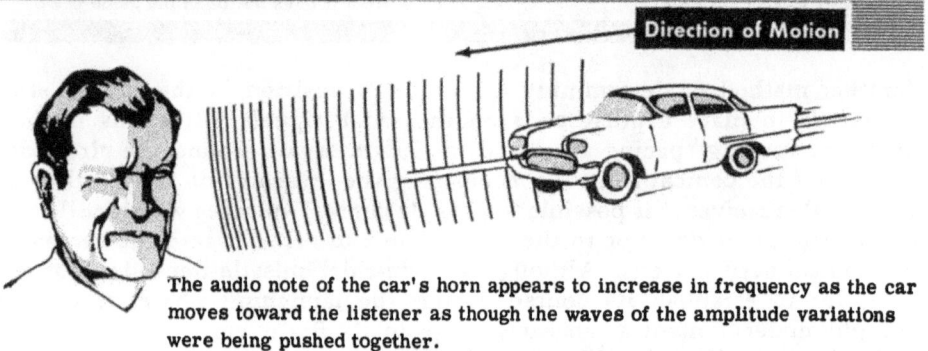

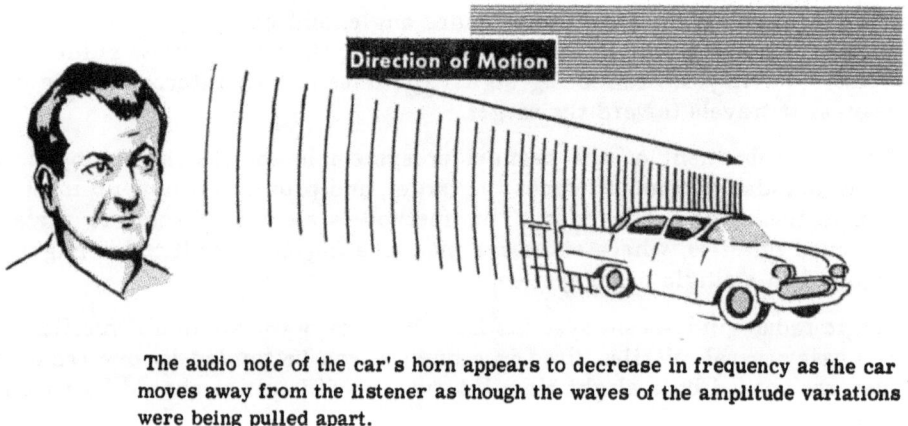

Direction of Motion

The audio note of the car's horn appears to increase in frequency as the car moves toward the listener as though the waves of the amplitude variations were being pushed together.

Direction of Motion

The audio note of the car's horn appears to decrease in frequency as the car moves away from the listener as though the waves of the amplitude variations were being pulled apart.

Doppler Principles and Applications (contd.)

Using the *Wolmann Cutoff* method (named for Dr. Wolmann of Dresden) the Doppler effect was employed to measure the velocity of the V-2 missile and to determine the point in the trajectory at which its rocket thrust should be throttled back and cut off.

Using the code words of that era the operation of the *Wolmann* system may be described as follows: A ground-transmitter, *Neapel I*, emitted a signal of frequency f_0. A receiver in the missile, *Ortler*, received the frequency $f_0 - \Delta f$ (Δf being the change due to the Doppler shift). To avoid interference with the *fundamental* transmission frequency, and improve the accuracy of the system, the frequency was doubled at the missile before sending it back to the ground evaluation center, *Capri*. Of course it was received back at that point with a further Doppler shift to provide a total signal-frequency of $2 f_0 - 4 \Delta f$ for comparison with the second harmonic, $2 f_0$.

When the frequency shift indicated the proper velocity for fuel cutoff at V-2, a command-signal was sent through the radio transmitter, *Palermo*, to a radio receiver, *Honnef*, in the missile to produce the desired reduction in thrust, which was accomplished in two stages; first, a cutback from 25 tons to 8 tons thrust and second, a complete thrust cutoff 3 seconds later. This method was just one of the techniques employed for controlling the fuel cutoff of the V-2. Another method used more often involved self-contained integration of acceleration to determine the proper point to reduce the thrust.

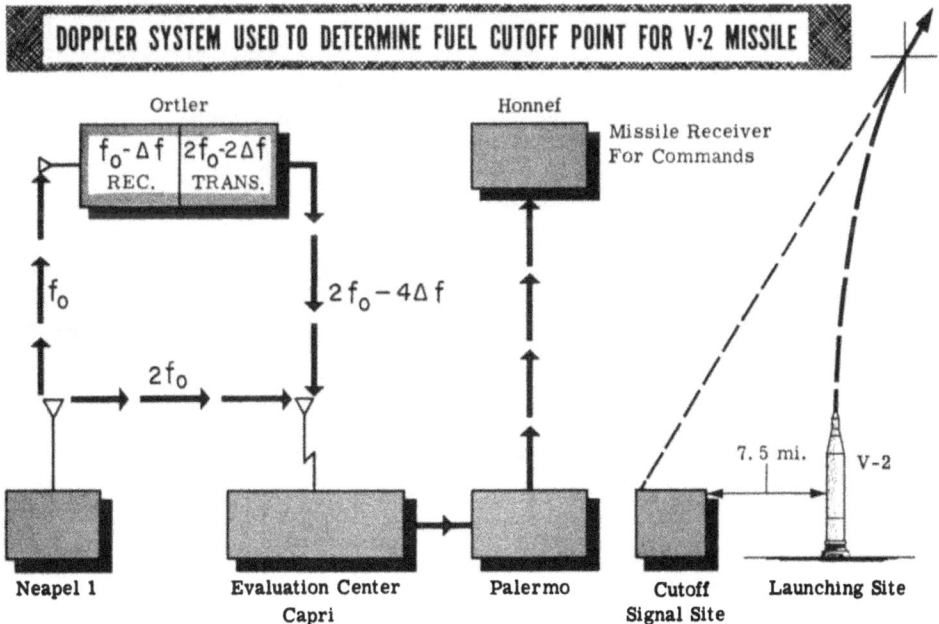

DOPPLER SYSTEM USED TO DETERMINE FUEL CUTOFF POINT FOR V-2 MISSILE

Doppler Principles and Applications (contd.)

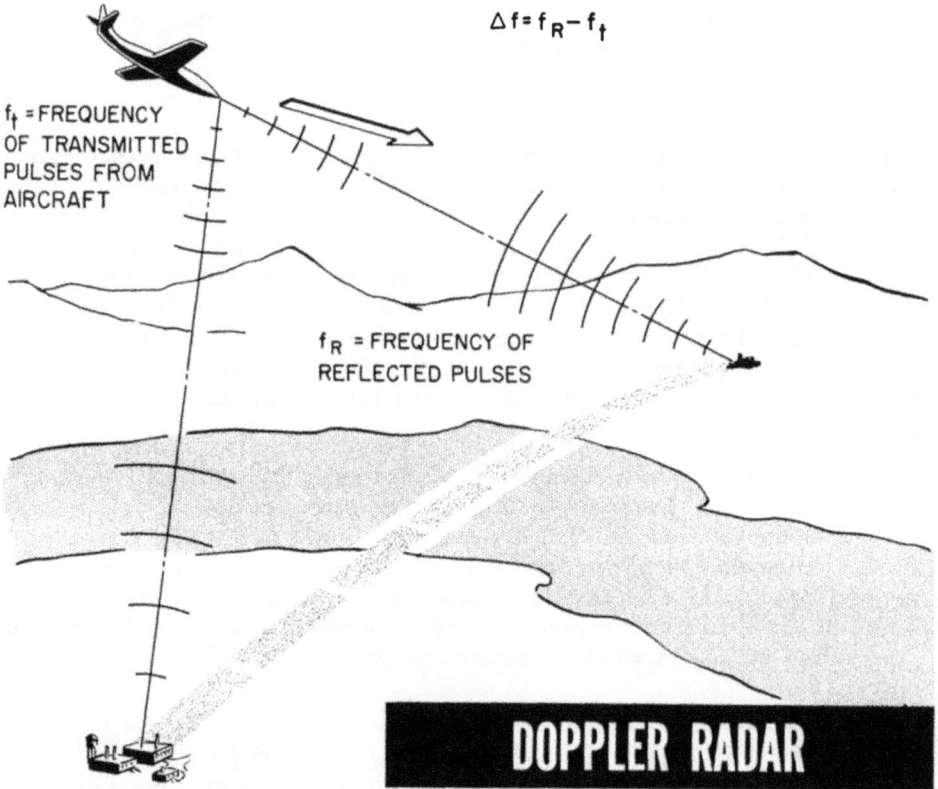

$\Delta f = f_R - f_t$

f_t = FREQUENCY OF TRANSMITTED PULSES FROM AIRCRAFT

f_R = FREQUENCY OF REFLECTED PULSES

DOPPLER RADAR

Since that time applications of the Doppler principle have been developed for altimeters, airborne-radar systems, and homing heads. In these applications reflected waves are compared with the incident waves with regard to frequency shift. Since reflected signals are utilized for comparison purposes, all of these systems are considered to be in the radar family. The mathematics for determining the shift frequency is fairly simple. If an object at A radiates a signal of frequency f_t and if the signal is reflected from a moving object B, returning to A for comparison, the formula for the frequency shift is:

$$\Delta f = \frac{2V}{c} \times f_t$$

where c is the velocity of propagation of the electromagnetic signal and V is the velocity of B with respect to A. If A is moving relative to B at an angle, then the above quantity must be multiplied by the cosine of the angle to obtain the correct value.

Doppler Principles and Applications (contd.)

In the Doppler application to altimeters an FM technique is used to continually measure the frequency shift between the transmitted and reflected signals as received in the aircraft. Thus, even when the aircraft is moving at a constant distance with respect to the earth, the altitude will be indicated by the frequency comparison. Also, any change in altitude will be indicated in a Doppler shift of the frequency. A rather popular application of Doppler principle today is embodied in what is known as the *Doppler Navigator*. It may be applied to either aircraft or missiles. In its operation two beams are transmitted from the aircraft to the ground at the angle ϕ. As mentioned above the formula for Δf under these conditions would be:

$$\Delta f = \frac{2\,V_g}{c} \times f_t \times \text{cosine } \phi$$

where V_g is the velocity of the aircraft relative to the ground.

By using the two beams it is possible to determine the ground speed of the aircraft by the measurement of Δf and it is also possible to determine the drift angle from a desired heading by rotating the two-beam antenna system until equal Doppler frequencies are received from one beam and noting the required angle of rotation to achieve this equality. Drift angle can also be determined with a fixed two-beam system by measuring the Doppler-frequency shift associated with each antenna-beam and calculating the drift angle corresponding to such a difference in Doppler shift.

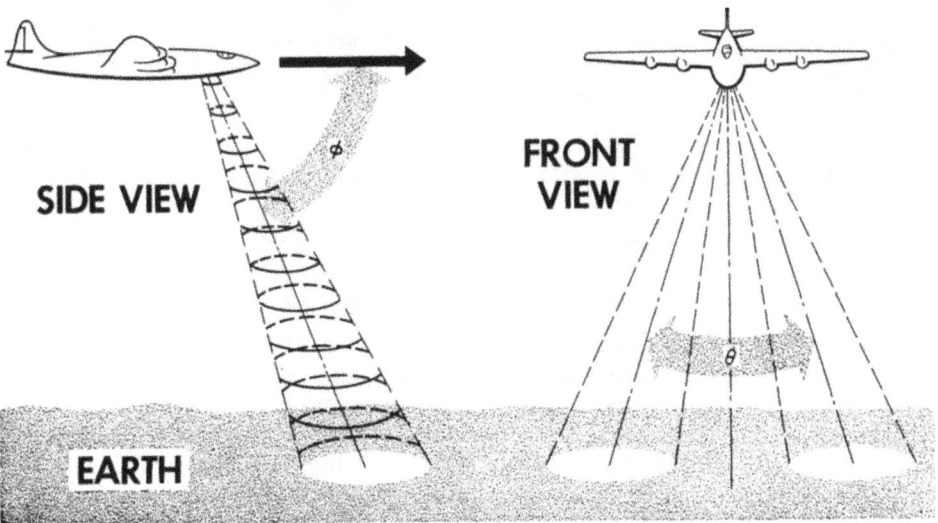

DOPPLER APPLICATION TO NAVIGATOR

SIDE VIEW

FRONT VIEW

EARTH

Missile Homing Methods

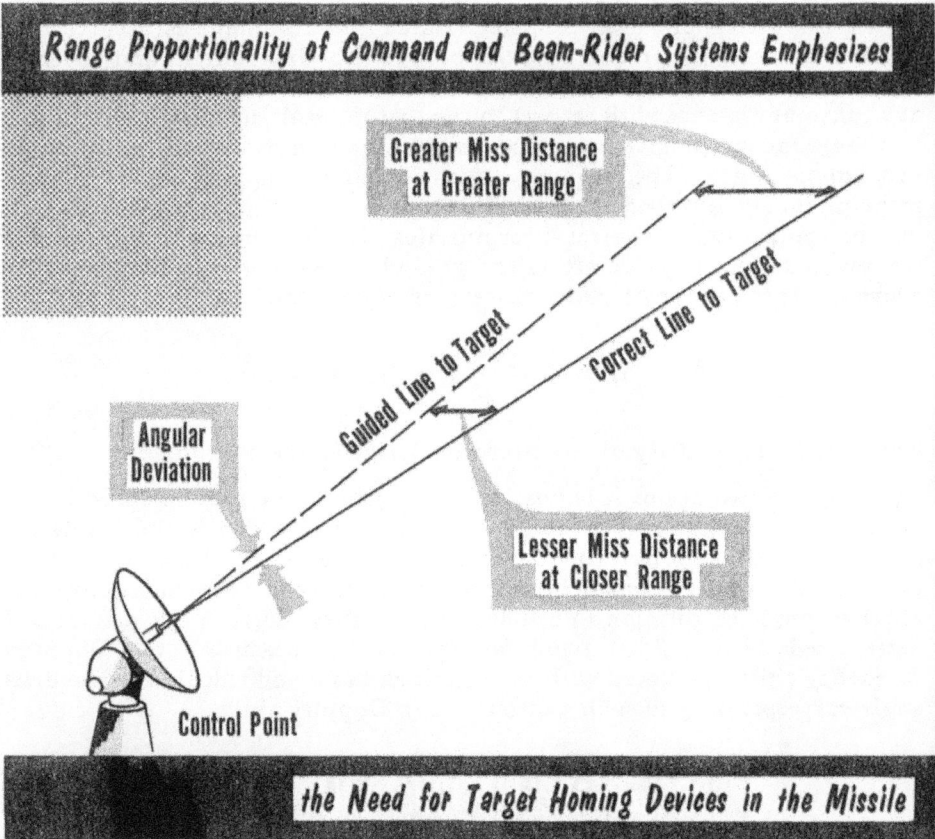

Range Proportionality of Command and Beam-Rider Systems Emphasizes

Greater Miss Distance at Greater Range

Correct Line to Target

Guided Line to Target

Angular Deviation

Lesser Miss Distance at Closer Range

Control Point

the Need for Target Homing Devices in the Missile

Doppler systems may be operated with either continuous-wave (CW) modulation or with forms of pulse modulation. The type of modulation will depend somewhat on whether short- or long-range operation is desired.

When an air-defense missile approaches the target its ability to achieve an effective contact can be greatly enhanced by the addition of homing techniques. This situation is due in a large measure to the fact that both command and beam-rider guidance systems are subject to angular inaccuracies which make them range-proportional; that is, as the range increases the missile error relative to the target in distance also increases.

As the missile travels away from a ground-control point the deviation from the correct line to the target is proportional to the distance from the control point for any given angular deviation. By adding a target-homing guidance in the missile it is possible to reverse the situation so that as the missile approaches the target, from which it is deriving the homing signal, the angular deviation results in a decreasing distance-deviation.

Missile Homing Methods (contd.)

Although they used practically none of them, due to their deteriorating general conditions in the latter part of World War II, the Germans had under development a number of homing systems for surface-to-air, air-to-surface, and air-to-air missiles. They investigated electromagnetic, optical, infrared, and acoustical devices. However, most of their effort was concentrated on electromagnetic and infrared types. They felt that homing devices were the right approach for guiding a missile during the last phase of its flight path. Although they had under development a number of homing systems, which we shall describe briefly, they were farther along with infrared homing-development near the end of the war.

Earlier we mentioned three classes of homing systems — active, semi-active, and passive types. It will be recalled that with the active method, a signal is radiated from the missile and a reflected signal is received in return at the missile for determining the correct action to obtain a homing course to the target.

In the semi-active type the target is *illuminated* by a radar external to the missile which contains a receiver for detecting the reflected signal from the target. In the passive type radiated energy from the target is used as a source for exciting the receiver at the missile, causing it to take a path toward the target. This energy might be in the form of radiation from a radar in an attacking aircraft, from the heat of its engines, from the noise of its motors or from its light image.

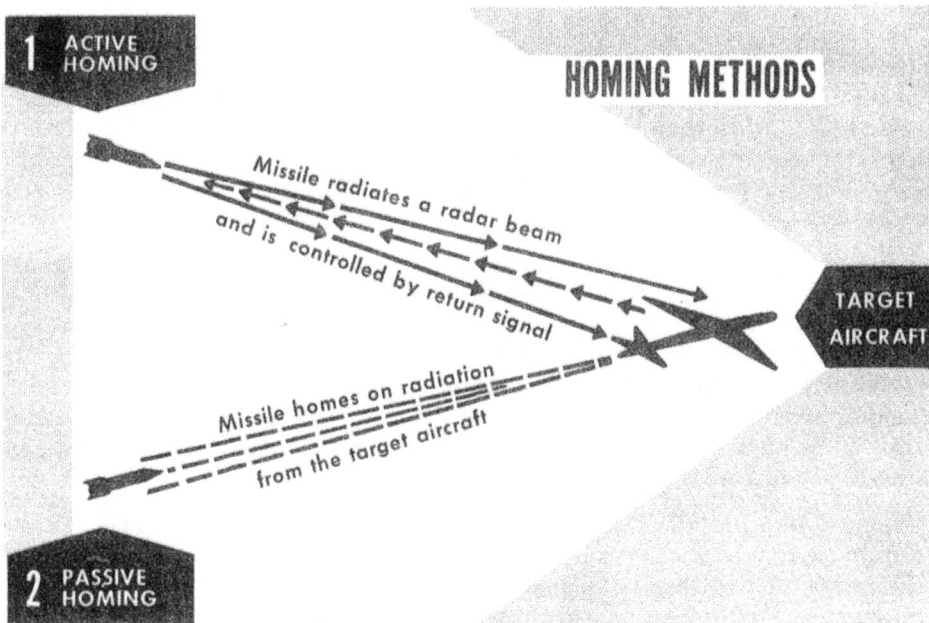

ACTIVE HOMING

HOMING METHODS

Missile radiates a radar beam and is controlled by return signal

TARGET AIRCRAFT

Missile homes on radiation from the target aircraft

PASSIVE HOMING

Missile Homing Methods (contd.)

The following details on German equipment under development at the end of World War II — *Max A, Max P*, and the *Moritz* systems — illustrate the principles of active, passive and semi-active systems:

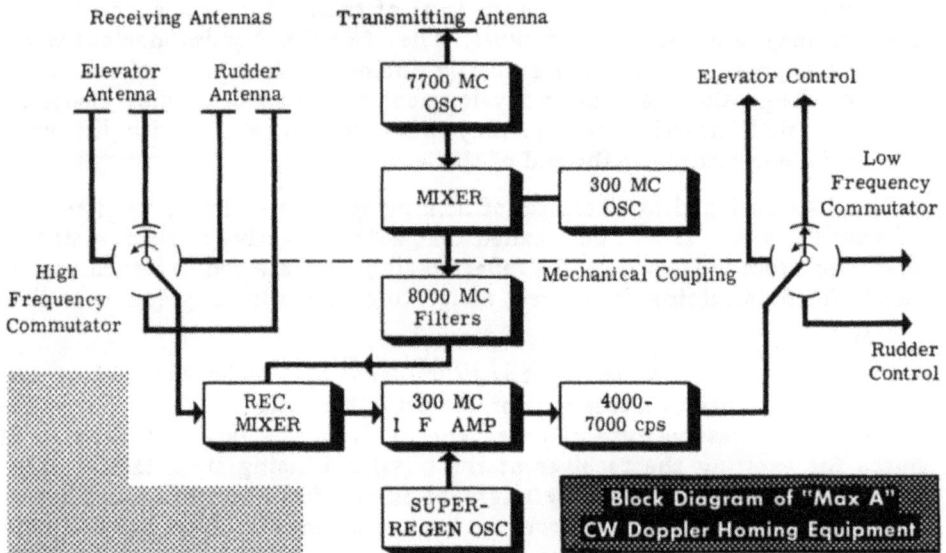

Block Diagram of "Max A" CW Doppler Homing Equipment

The transmitter of *Max A* radiated approximately 5 watts of CW power from a horn antenna at 7700 mc. This signal was obtained from an 8-slot magnetron working with a plate voltage of 600 volts and a field intensity of 1000 gauss. Since the received signal from the target would be shifted only a few thousand cycles away from the transmitted frequency due to the Doppler effect, it was necessary to eliminate the frequency-drift effects in the receiver.

A high order of stability was realized by beating a voltage from the transmitter with a frequency-stabilized oscillator at 300 mc within the Max A equipment. By selecting the sum frequency through a filter system, a local oscillator frequency of 8000 mc was available at the receiver to beat with the reflected signal from the target. This mixing action produced an intermediate frequency of 300 mc from which the 4000 to 7000-cycle Doppler-shift audio signal was detected to provide control signals to the elevator and rudder of the homing missile.

Designs were provided for both fixed and movable antennas. The voltage difference of the rudder and elevator-antenna sections was used to correct the course of the missile until the signal from the target was in the plane of symmetry of both the horizontal and vertical arrays.

Missile Homing Methods (contd.)

The receiver antenna was constructed of dielectric rods with a half-lobe angle of 15°. In the case of a 1° displacement of the target, the deviation of the receiver antenna from its 15° mean position produced sufficient change in the receiver output to operate the differential relays and to change the course of the missile.

In order to produce better decoupling, the transmitter antenna was constructed as a horn of sheet metal. Two antenna arrays were employed for different homing approaches: a rigid antenna head, fixed on the axis of the missile for pursuit courses, and a movable antenna for collision courses. The pursuit course was planned for air-to-air operations, and the collision course for ground-to-air use. When operating with the collision-course system, it was necessary to guide the missile from the ground until it came within the range of its own self-contained homing device.

After locating the target, the seeking head in the first stage of the array automatically turned itself toward the target without affecting the navigation of the missile. Upon reaching a distance of approximately 3000 feet from the target, the angle between the axes of the missile and the antenna head was fixed, and automatic navigation was begun with the aid of a potentiometer. The head turned in the direction of the target, moved the potentiometer, and disturbed the balance of a resistance bridge, which altered the course of the missile until the fixed angle between the array head and the missile axis was obtained.

The movement of the antenna was produced by two course-setting motors — one moved the antenna drive directly, and the other produced the rectangular movement through a flexible shaft. Thus it was possible to set the antenna to any direction in space.

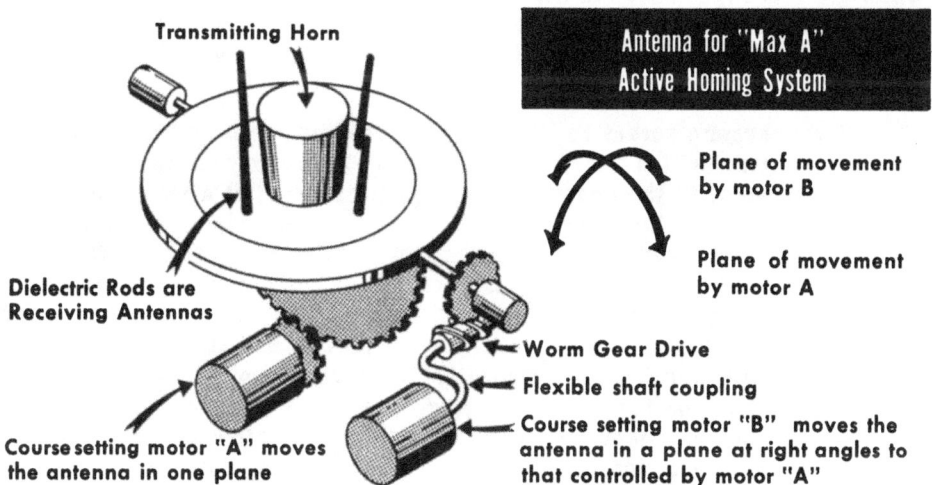

Transmitting Horn

Antenna for "Max A"
Active Homing System

Plane of movement by motor B

Plane of movement by motor A

Dielectric Rods are Receiving Antennas

Worm Gear Drive

Flexible shaft coupling

Course setting motor "A" moves the antenna in one plane

Course setting motor "B" moves the antenna in a plane at right angles to that controlled by motor "A"

Missile Homing Methods (contd.)

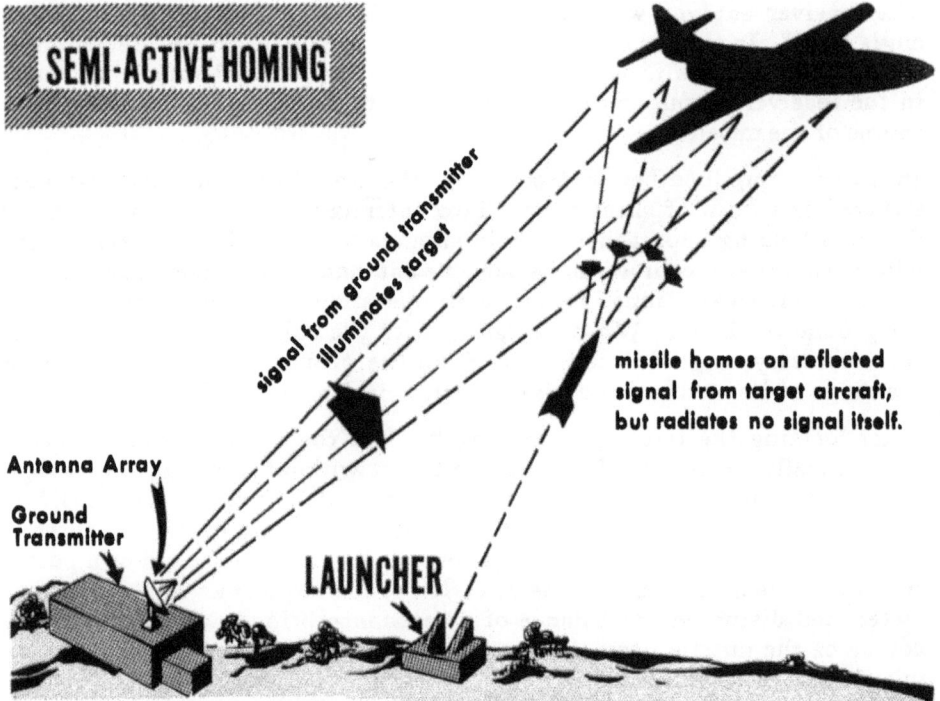

SEMI-ACTIVE HOMING

signal from ground transmitter illuminates target

missile homes on reflected signal from target aircraft, but radiates no signal itself.

Antenna Array

Ground Transmitter

LAUNCHER

The *Max P* passive homing system was constructionally similar to the *Max A* active homer, except that the transmitter was not included. It homed on the radar emissions of hostile aircraft and it is said that ranges of up to 30 miles were obtained against 3-cm airborne radars. However, a major difficulty was experienced in the long time interval of 2 to 3 seconds between the radar signal sweeps across the missile. Because of this no homing information was available to this passive system during relatively long intervals of time.

The *Moritz* system serves to illustrate the semi-active homing application. The transmitter, which had a CW output of 10 watts, was located on the ground instead of in the missile. Although it operated at a somewhat lower frequency (430 mc), than *Max A* or *Max P*, it made use of the Doppler effect between the direct wave and the reflected wave and navigated on the latter. This system had advantages of small weight and low cost in so far as the missile was concerned, since the transmitter was located on the ground, and no need for direct coupling between the transmitter and the receiver. However, the missile could, unless carefully launched, attack the ground transmitter instead of the target. Also the ground transmitter could be located and knocked out by enemy action. The details of present-day homing systems for missiles are classified by military security regulations.

Infrared Homing Techniques

World War II saw the start of the initial development of infrared homing systems for missiles to be launched against targets with heat radiation as the detectable quality. In Germany basic laboratory investigations to determine the feasability of using infrared radiations for military purposes started in the early 1930s. Both active and passive systems appeared to be feasible, but the simplicity of the passive systems presented the greatest attraction. The heat radiation from the exhaust gas of reciprocating engines appeared adequate, and without the need of any active means for *illuminating* the target. With the advent of jet engines and rockets the heat radiation from such an attacking aircraft or missile was further enhanced.

Portion of Electromagnetic Spectrum Showing Regions Used for Radar and Infrared Homing

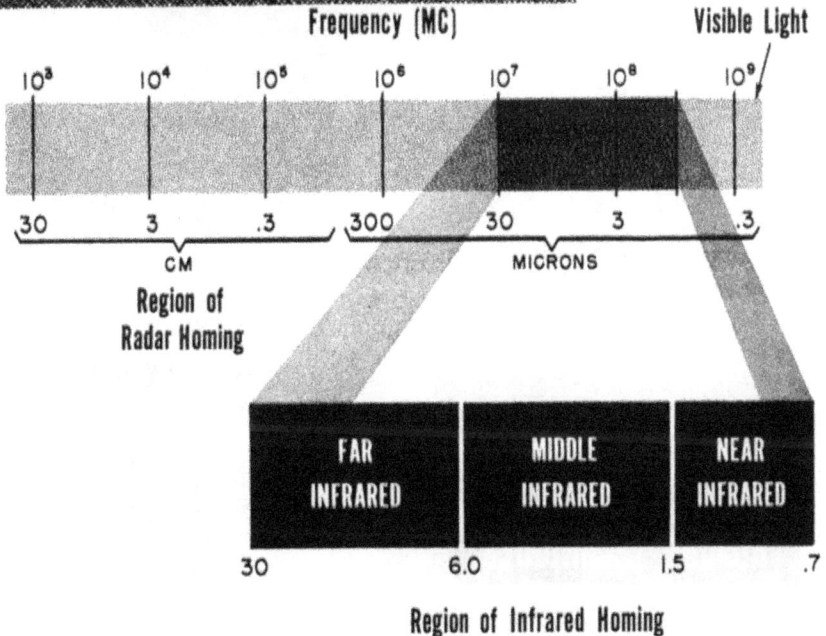

Region of Infrared Homing

The micron, equal to 1 millionth of a meter or 1/10,000 of a centimeter, is used to express wavelength in the infrared region. We have seen how radar homing can be employed in the centimeter range of radiation. Infrared homing is generally employed in a limited range of the infrared spectrum from about 1.5 to 6 microns.

Infrared Homing Techniques (contd.)

In the region from about 0.8 to 1.5 microns most objects radiate less energy, but detectors which are capable of photographing the image are available. Below 0.8 micron we find visible light, ranging from about 0.75 to 0.4 micron from the infrared through ultraviolet. Radiation from aircraft engines, either reciprocating or jet types, appears primarily in the range of 2 to 5 microns.

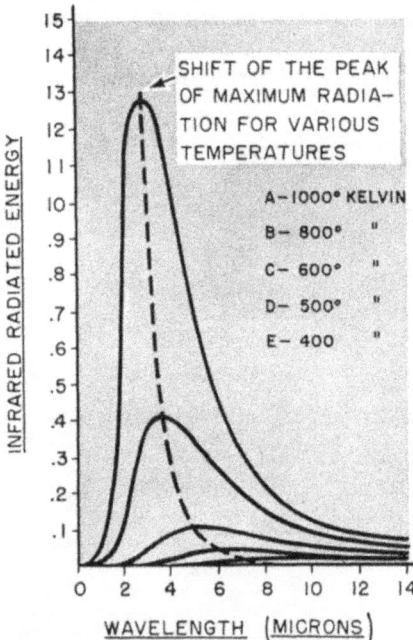

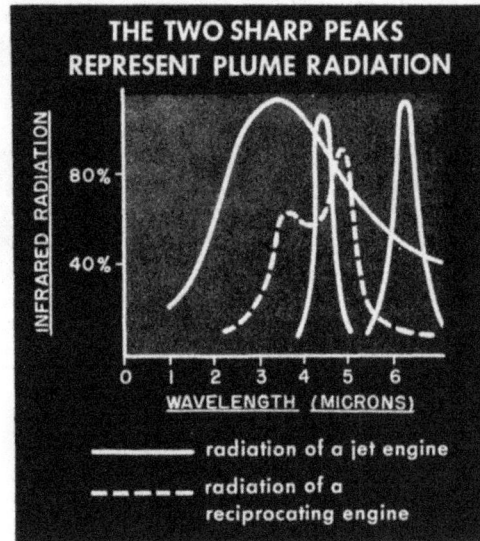

SHIFT OF THE PEAK OF MAXIMUM RADIATION FOR VARIOUS TEMPERATURES

A—1000° KELVIN
B— 800° "
C— 600° "
D— 500° "
E— 400 "

INFRARED RADIATED ENERGY

WAVELENGTH (MICRONS)

THE TWO SHARP PEAKS REPRESENT PLUME RADIATION

INFRARED RADIATION

WAVELENGTH (MICRONS)

——— radiation of a jet engine

- - - - radiation of a reciprocating engine

INFRARED RADIATION APPLIED TO HOMING TECHNIQUES

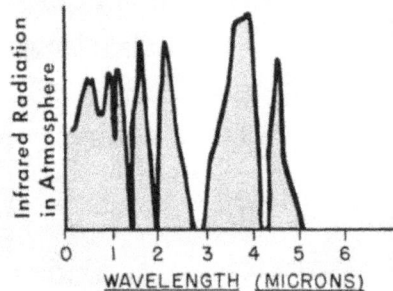

Infrared Radiation in Atmosphere

WAVELENGTH (MICRONS)

In the search for a suitable detector of radiation in the range of 3 to 5 microns, the photoconductivity of lead sulphide was found suitable in 1932. Practically all infrared devices developed in World War II utilized lead-sulphide cells. However, by 1944 lead-selenide photoconductive cells were also ready for laboratory use. Since that time lead-telluride, germanium-and indium-antimonide have been considered as detectors in this range of radiation.

Infrared Homing Techniques (contd.)

In order to increase the sensitivity of an infrared homing system the detector is placed at the output of an optical system of the *Schmidt* type. Since a-c signals can be amplified more easily than d-c, the scanner in front of the infrared detector cell is used to modulate the input signal. The parabolic mirror, which is usually about 3 inches in diameter, spins on an eccentric axis giving a reception pattern somewhat similar to a conical scan-radar. The whole optical system may be arranged to be movable about its axis so that it can rotate into the direction of the maximum heat signal. This motion is transferred by electrical means to a servosystem which can vary the control surfaces to bring the missiles into the correct course.

Infrared systems suffer from rather severe attenuation factors in the atmosphere, the optical system and the detector-response characteristics. For instance, in a typical situation only about 3% of the available heat energy at the source may be available for useful application in the system. Although the transmissivity of infrared radiation has been found to be greater than that of visible light under all weather conditions, even with fog or through clouds it is quite small under the latter conditions. Despite these limitations, the use of infrared homing equipment for certain applications, such as air-to-air missiles has been found to be quite effective.

THE ENTIRE OPTICAL SYSTEM MAY BE ARRANGED TO BE MOVABLE ABOUT ITS AXIS

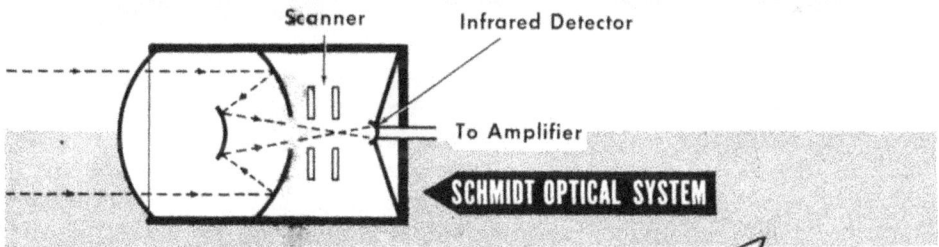

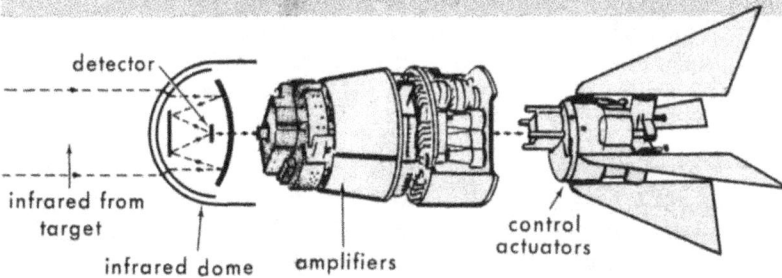

Scanner

Infrared Detector

To Amplifier

SCHMIDT OPTICAL SYSTEM

detector

infrared from target

infrared dome

amplifiers

control actuators

INFRARED GUIDANCE SYSTEM

Proximity Fuzing

SHADED AREA covers the region in which sufficient reflected signal is available from the target to actuate the proximity fuze.

MISSILE CARRYING PROXIMITY FUZE can receive sufficient reflected signal from the target to detonate warhead as soon as it approaches the shaded area. Thus, it DOES NOT NEED TO CONTACT THE TARGET TO DESTROY IT.

Proximity Fuzing of a Missile Can Increase the Kill Probability

Even with accurate command or beam-guidance and a homing system it may not be feasible to obtain enough direct hits against high–speed targets to permit the utilization of contact fuzes for detonation of the warhead. Also, it is not practical to use a mechanical preset fuze. Thus, proximity fuzes are often essential to the destruction of a high percentage of the targets. A variety of proximity-fuzing arrangements were developed during World War II. From this side of the World, their application to antiaircraft shells was the major operational feat. However, the Germans did extensive development work on a number of systems which they had hoped to use in both defensive and offensive missiles.

Most of the U.S.A. antiaircraft shell fuzes operated on the basis of a Doppler-frequency shift between a radiated and reflected signal. The same oscillator, which provided the transmitted signal, also served as a beat-frequency source for converting the reflected signal to an a-f signal. This signal was amplified further to fire a small thyratron tube controlling the squib in the warhead.

Proximity Fusing (contd.)

A typical German missile fuze development was conducted in conjunction with the *Max A* homing equipment described earlier. In this arrangement some of the voltage from the 7700 mc oscillator was fed through a coupling resistance to the input of a relatively simple receiving channel, which also picked up reflected signals from the target. The local oscillator signal of 7700 mc beat with the reflected signal to give an audio output ranging from 2 kc to 5 kc, which was selectively amplified. The detected output was used to operate a relay, which controlled the firing of the warhead.

A TYPICAL GERMAN MISSILE FUZE DEVELOPMENT

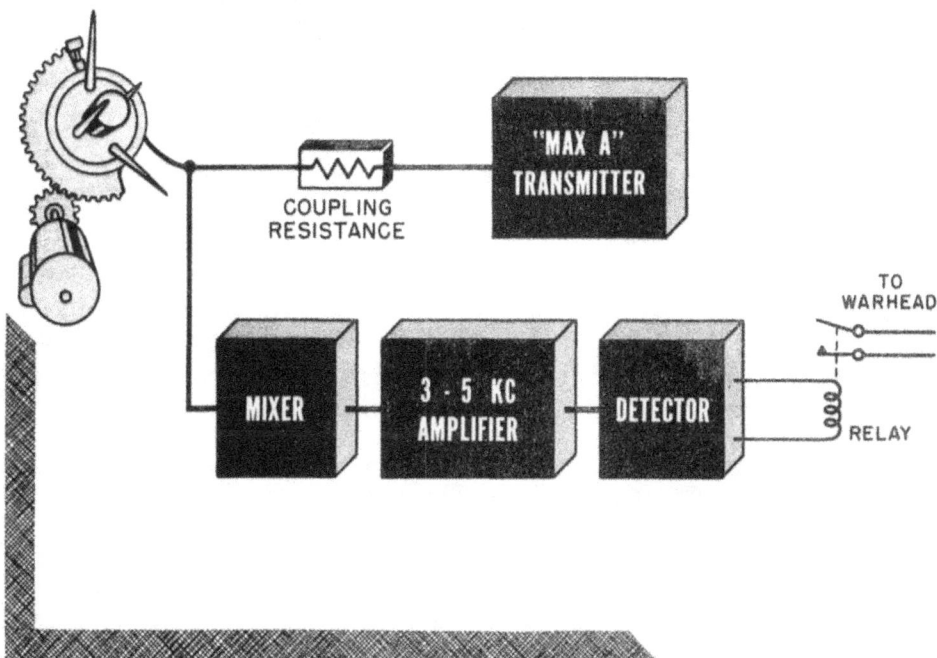

Several other types of fuzes, utilizing Doppler effects, frequency modulation, antenna-interference variations, electrostatic fields, magnetic-field effects, acoustical and optical qualities of the target were in various stages of development at the end of World War II. An analysis of all of these types by one of the German developers indicated that the most promise probably lies at the very highest radio-frequencies and at the longer wavelengths of infrared radiation.

Introduction to Gyroscopes

Ordinary tops and small toy gyroscopes are generally driven by one pull of a string to bring their mass up to a velocity which keeps them in stable motion for a short time. These commonplace examples utilize the same fundamental principles as those of the much more elaborate and precise gyroscopes so essential to present-day navigation. The top is a good example of simple gyroscopic action. However, it is not suspended in any sort of frame. The toy gyroscope, available in almost every part of the world, is suspended in a frame, and is an example of a single-axis type.

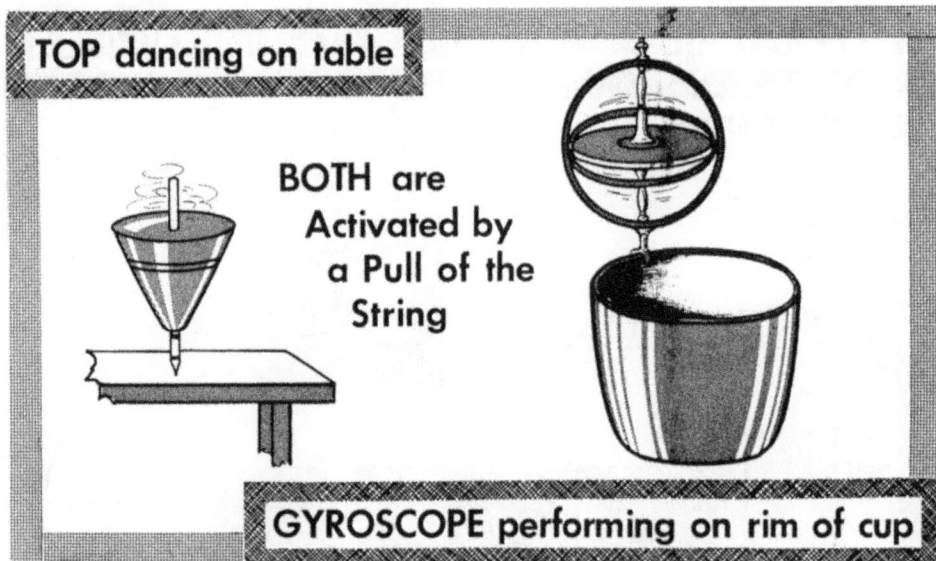

TOP dancing on table

BOTH are Activated by a Pull of the String

GYROSCOPE performing on rim of cup

Ever since the invention of the wheel, man has witnessed gyroscopic action to some degree in this familiar form. The invention of the top is credited to the Chinese many centuries ago. Leon Foucault, a French scientist, is acknowledged as the inventor of the gyroscope — as we know it — in 1852. However, 20 years earlier a German, Herr Bohnenburger, made similar experiments. At that time a Scot, Edward Sarg, suggested an experiment similar to that performed by Foucault.

In his experiment Foucault utilized the fact that the gyro axis points at a fixed direction in space. In accordance with one of Newton's laws of motion the spinning rotor resists changes in its direction of motion. In Foucault's experiment a scale mounted around and separate from the gyro frame followed the rotation of the Earth, while the gyro axis pointed at a fixed direction in space thus indicating the Earth's relative motion. The gyroscope even received its name from this application (*gyro* for revolution and *scope* for observer; thus, it is literally a revolution observer).

GYROSCOPIC FUNDAMENTALS

The Planet Earth as a Gyroscope

Actually, gyroscopic action is as old as the universe. Ever since matter started rotating about a point in space gyroscopic principles were in action. In fact, we spend most of our lives on a gyroscope, the Earth. Within the knowledge of man its axis has pointed toward the North Star (Polaris) generally. As we all know, it rotates at 1 revolution per 24 hours, and travels around the Sun in an elliptical orbit at 1 revolution per year. However, due to its mass it develops the substantial angular momentum of 6.5×10^{31} tons ft^2 per sec.

The fact that the tilted poles of the Earth are pointed in a relatively fixed direction due to its gyroscopic behavior accounts for the four seasons as it rotates about the Sun at more than 60,000 miles per hour. Even than a 47° deviation of the Earth's axis takes place every 12,900 years due to an effect known as precession and causes the seasons to vary somewhat over this rather long period.

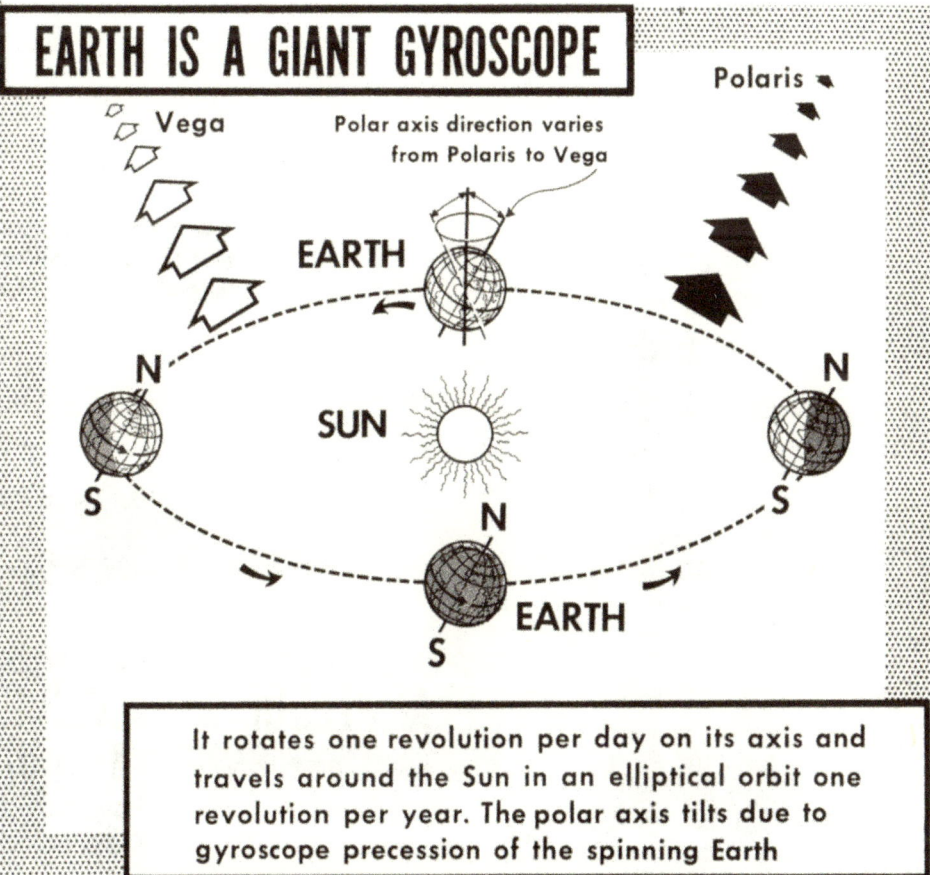

EARTH IS A GIANT GYROSCOPE

Polaris

Vega

Polar axis direction varies from Polaris to Vega

EARTH

N
S

SUN

N
S

N
EARTH
S

It rotates one revolution per day on its axis and travels around the Sun in an elliptical orbit one revolution per year. The polar axis tilts due to gyroscope precession of the spinning Earth

GYROSCOPIC FUNDAMENTALS

Other Gyroscopic Examples

Other rather common gyroscopic examples are found in all types of spinning and rotating bodies. For instance, gun barrels are rifled to give shells a spinning effect for stabilization. Simple rockets are arranged with jets exhausting in the proper direction to rotate them for the sake of stability in their trajectories. Of course, spinning objects in the air can produce other than the perfectly directed courses attainable in a vacuum. Spinning objects revolving against an air cushion will tend to curve rather than follow a straight line. However, if allowances are made for this effect, the spinning or rotating action will still provide stability in a trajectory.

Gyroscopic stabilization may be realized by spinning the object to be controlled; also, by utilizing a gyro rotor of such large mass that it dominates the stability aspect when rotated at the required velocity; or by incorporating one or more small gyroscopes inside of the object and utilizing information derived from it to keep the object on a desired course. One of the most recent applications of rotating a rapidly moving mass to give it stability is found in the spinning of satellites.

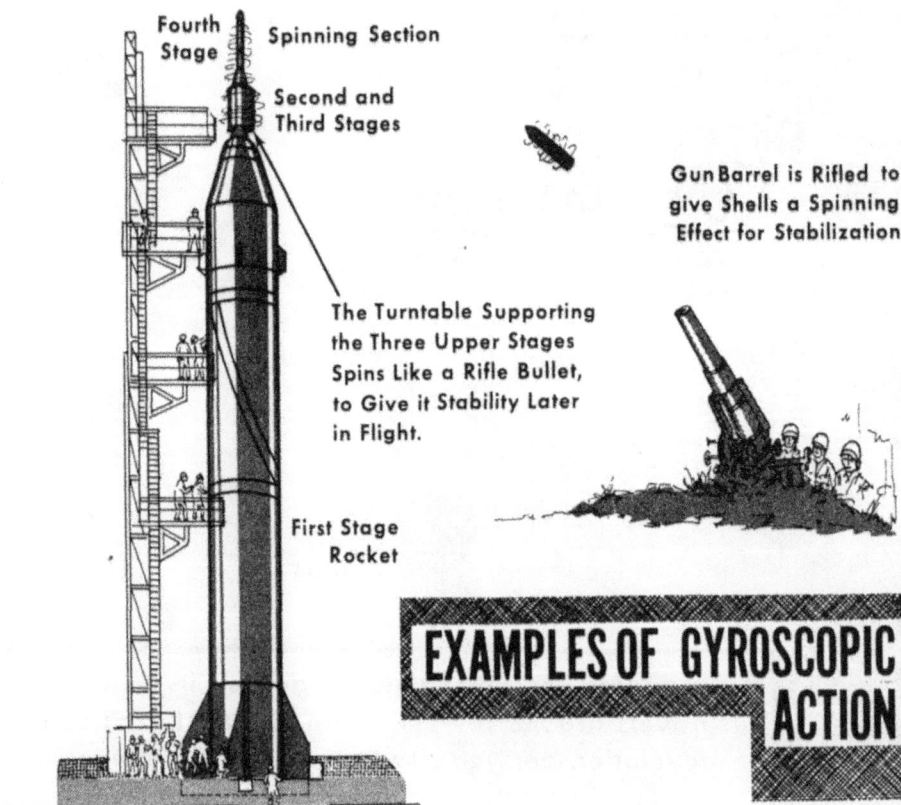

Fourth Stage
Spinning Section
Second and Third Stages
Gun Barrel is Rifled to give Shells a Spinning Effect for Stabilization
The Turntable Supporting the Three Upper Stages Spins Like a Rifle Bullet, to Give it Stability Later in Flight.
First Stage Rocket

EXAMPLES OF GYROSCOPIC ACTION

Rigidity in Space

A GYRO'S RIGIDITY IN SPACE CAN BE EXPRESSED BY A SPIN VECTOR

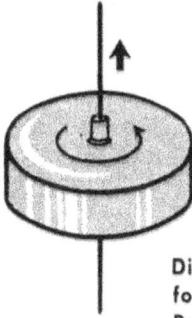

Direction of the Spin Vector for Counterclockwise Rotation of the Gyro Vector

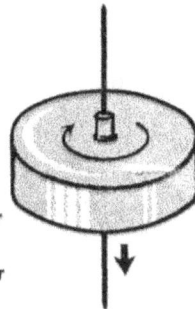

Direction of the Spin Vector for Clockwise Rotation of the Gyro-Rotor

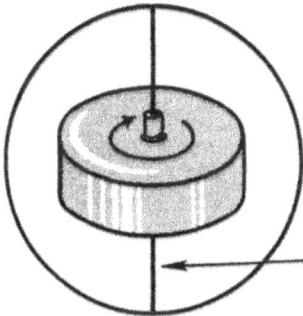

External Torque applied to a Gyro-Rotor acts to Disturb its Rigidity in Space. The Gyro Responds by Precessing.

One of the principal characteristics of a gyroscope — axial rigidity in space — has already been mentioned several times previously. This feature is its prime quality as a device for navigational purposes. Its fixed direction can be utilized for maintaining a desired heading, provided it can be associated with adequate references so that the rigidity of the rotor axis to external torques is meaningful to the navigator.

The gyroscopes previously mentioned have either been without a supporting ring about their rotational axis — as in the case of the top and the Earth — or with a single ring as in the case of the simple toy gyroscope. Also, one could mention the simple compass of the type used by Foucault in his experiment. However, navigational gyros are suspended in two or three rings or gimbals — as they are called — giving them an ability to respond to forces about the axes with which they are associated.

Precession

When a gyroscope responds to deviating forces about its axes, it displays another very important characteristic known as *precession*. Spinning tops wobble about their axes. Especially when they slow down, they allow external forces to produce more effect than when they are spinning fast. Also, the Earth slowly describes a significant axial deviation in space over a long period of time. However, to make precession useful for navigational purposes, it is necessary in a man-made gyroscope to provide one or two frames or gimbals in addition to the frame surrounding the spin axis.

In these frames the gyro realizes one or two degrees of freedom in which precession can take place. The effect of precession is simply such that when a force is applied at right angles to the axis of the rotating mass, the gimbal ring supporting the axis of the rotating mass will not move in the direction of the applied force but will rotate at right angles to it.

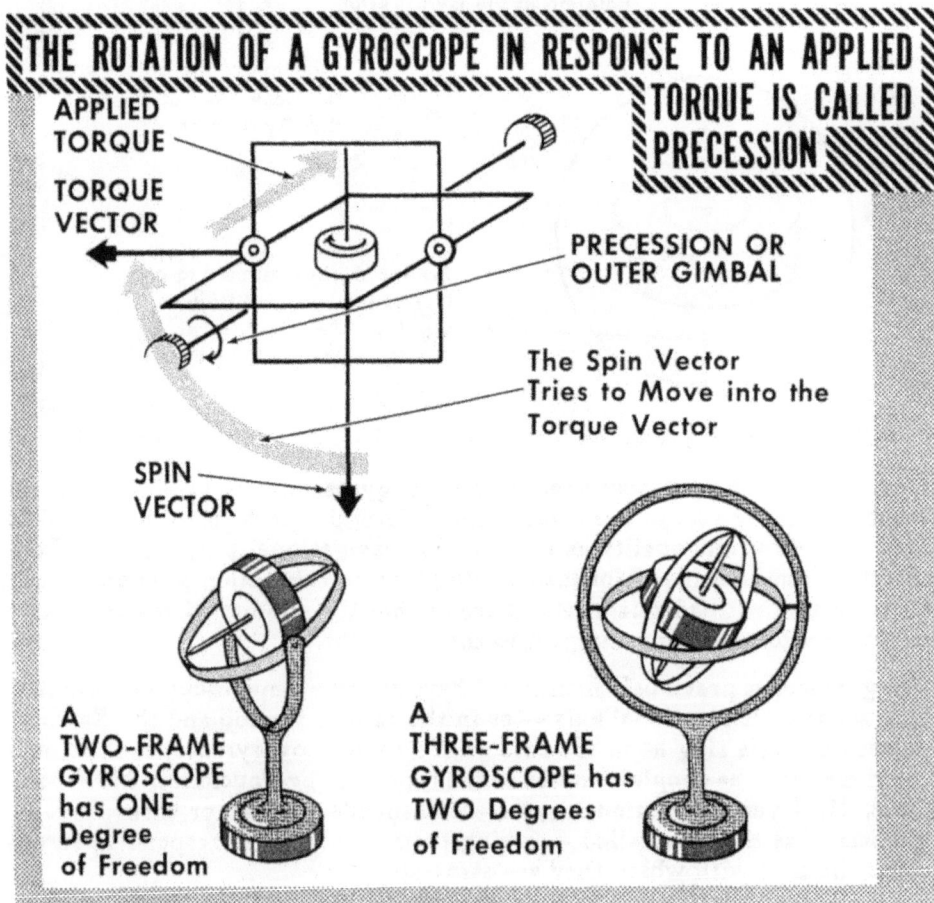

THE ROTATION OF A GYROSCOPE IN RESPONSE TO AN APPLIED TORQUE IS CALLED PRECESSION

APPLIED TORQUE

TORQUE VECTOR

PRECESSION OR OUTER GIMBAL

The Spin Vector Tries to Move into the Torque Vector

SPIN VECTOR

A TWO-FRAME GYROSCOPE has ONE Degree of Freedom

A THREE-FRAME GYROSCOPE has TWO Degrees of Freedom

GYROSCOPIC FUNDAMENTALS

Precession (contd.)

The precession of a rotating gyro wheel, when subjected to external torques can be readily explained through the use of vector diagrams. By considering the torques as being applied to small unit masses in a gyro rotor wheel, it is rather easily understood that the rotor shifts to align itself with the direction of the vector sum of its torque and that of the external torque.

A simple rule which may be applied to determine the direction of precession is that of pointing the forefinger of the right hand in the direction of the external torque as seen from the rotor axis, closing the other fingers in the direction of the rotation of the gyro rotor, and noting the direction in which the thumb points.

EXTERNAL TORQUE APPLIED AROUND AXIS A-A

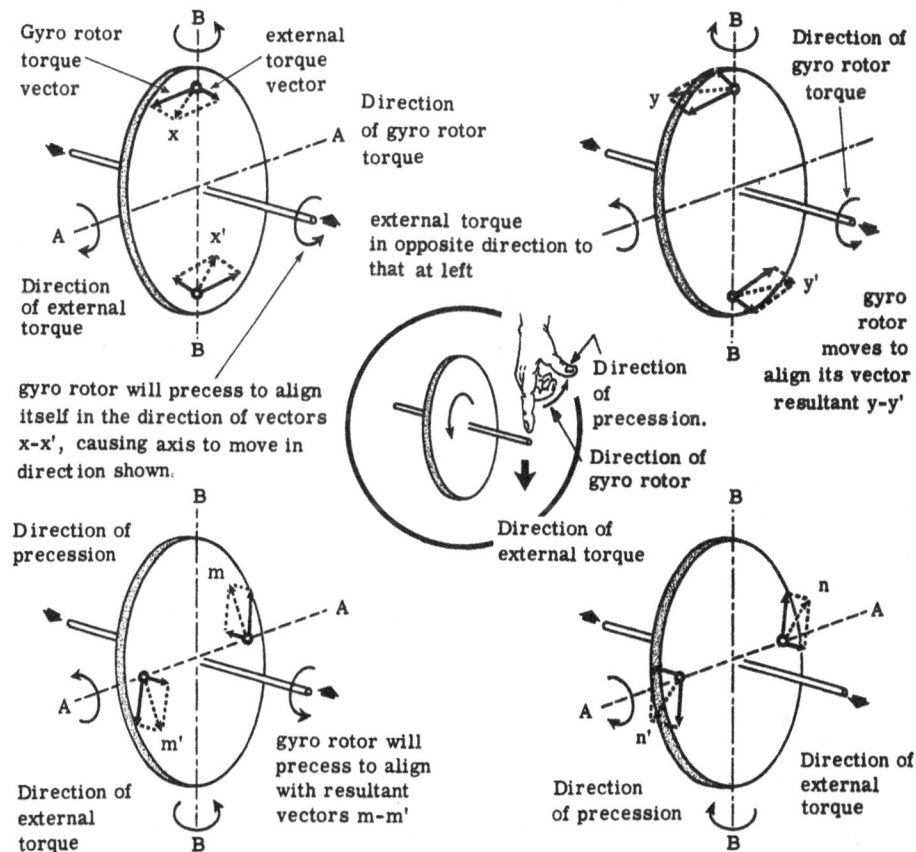

Gyro rotor torque vector

external torque vector

Direction of gyro rotor torque

Direction of gyro rotor torque

x

A

external torque in opposite direction to that at left

y

A

Direction of external torque

gyro rotor will precess to align itself in the direction of vectors x-x', causing axis to move in direction shown.

Direction of precession.

Direction of gyro rotor

gyro rotor moves to align its vector resultant y-y'

y'

Direction of precession

m

A

Direction of external torque

Direction of external torque

gyro rotor will precess to align with resultant vectors m-m'

Direction of precession

n

A

n'

Direction of external torque

EXTERNAL TORQUE APPLIED AROUND AXIS B-B

Precession (contd.)

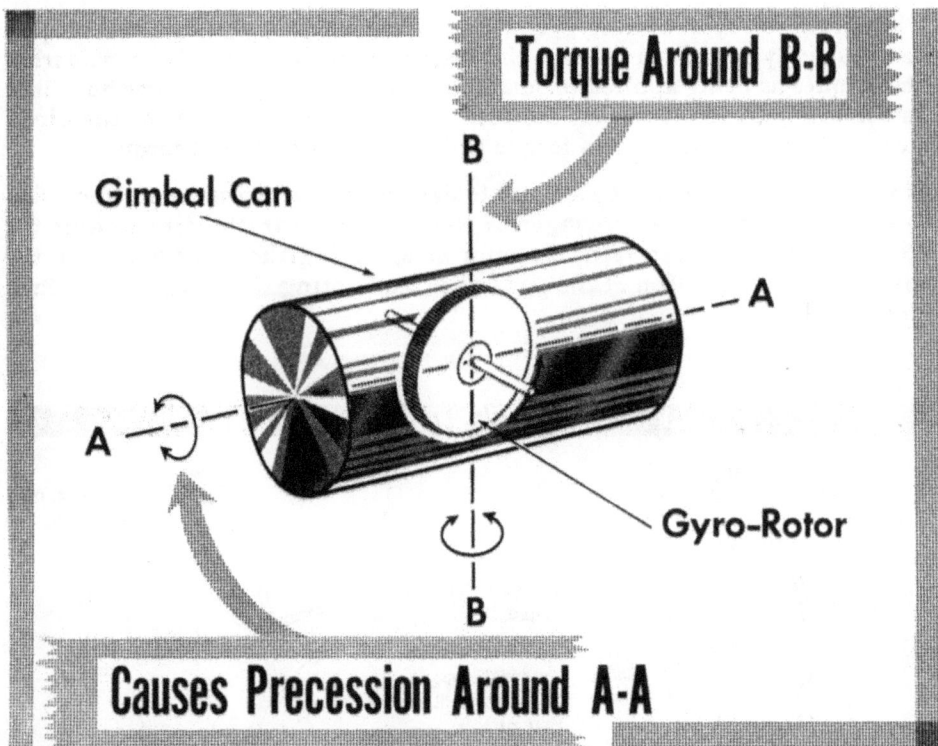

Torque Around B-B

Gimbal Can

A

A

Gyro–Rotor

Causes Precession Around A-A

From the diagrams on page 1-75, it is apparent that when a gyro rotor is sealed inside a gimbal can, variations in external torque around axis B-B will cause the gimbal-rotor assembly to precess about axis A-A. As shown on page 1-81, signals generated by the precession output are used to bring the gyro back to its original direction in space. The rate at which the gyro–rotor moves toward the new torque vector in precession is directly proportional to the applied torque, but indirectly proportional to the angular momentum of the gyro wheel about its own axis.

The basic relationship describing gyro-rotor behavior is expressed by the formula $\overline{W}_p = \overline{T}/\overline{H}$. $\overline{W}_p$ is the absolute precessional velocity of the spin axis (radians/second), $\overline{T}$ the applied torque (dyne-centimeters), and $\overline{H}$ the rotor angular momentum (gram-centimeters^2 per sec).

Since the gyro's precessional quality is largely dependent on $\overline{H}$, they are generally rated in terms of the rotor angular momentum. Gyroscopes are often referred to as 10^3, 10^5, or 10^6 units. These numerical quantities are a measure of $\overline{H}$, which is variously expressed as gram-centimeters^2 per second, dyne-centimeters-seconds per radian, or dyne-centimeters per radian per second — all of which are equivalent.

How the Gyroscope is Employed

The output of the gyro is proportional to the torques to which it may be subjected in its operational application. If it has only one gimbal around which precession can be measured, it can respond only to torques in one direction, and is said to have *one degree of freedom*. One can study the manner in which precession signals from a single gimbal gyro may be utilized by considering that the gimbal can is mounted on a movable platform rather than an aircraft or missile frame.

With the application of external torque about the platform axis both it and the attached gyro will start to move to a new position depending upon the amount and direction of the torque. Simultaneously a signal will be generated due to the gyro precession. This voltage can be picked off and fed through an amplifier. Then it is fed back through a motor which restores the platform and the attached gyro to their original positions. The platform to which the gyro container is attached may be the body of an aircraft or missile which would be restored to its original position by gyro-stabilization. Since the single-degree-of-freedom gyro can precess in only one plane of motion, an additional gyro is required for each additional control axis.

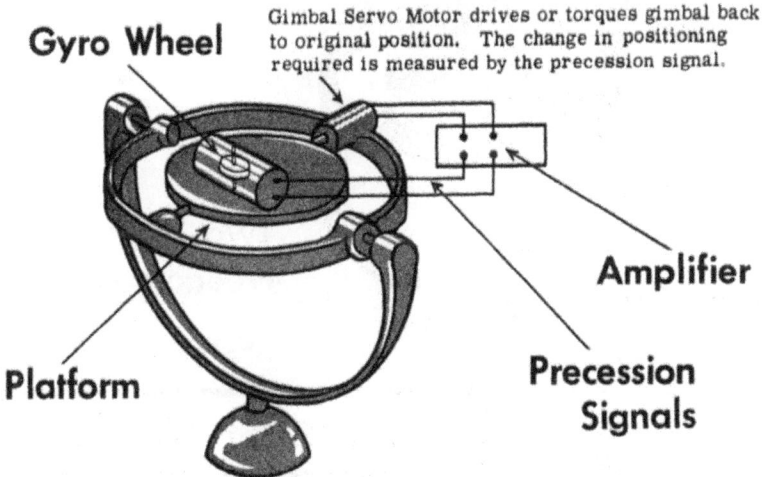

HOW PRECESSION IS EMPLOYED

Gyro Wheel

Gimbal Servo Motor drives or torques gimbal back to original position. The change in positioning required is measured by the precession signal.

Amplifier

Platform

Precession Signals

Single-Degree-of-Freedom Stabilization

How the Gyroscope is Employed (contd.)

If a gyro has two gimbals around which precessions can be measured, it can respond to torques in two directions, and is said to have *two degrees of freedom*. For this type of unit two separate signals can be made available, and can be applied to two platform gimbals to restore them to their positions when fed through suitable servo-loops. Thus, it is possible to obtain signals from one gyro unit to correct for any two of three possible deviations — roll, pitch and yaw.

A single-degree-of-freedom gyro will provide for correcting control in only one of these frames. Both types of gyros have been employed in a variety of applications. Some prefer the single-degree-of-freedom type because it is simpler to build and hold to tolerances in manufacturing, while others prefer the two-degree type because two units instead of three will suffice to control a system in three planes of motion.

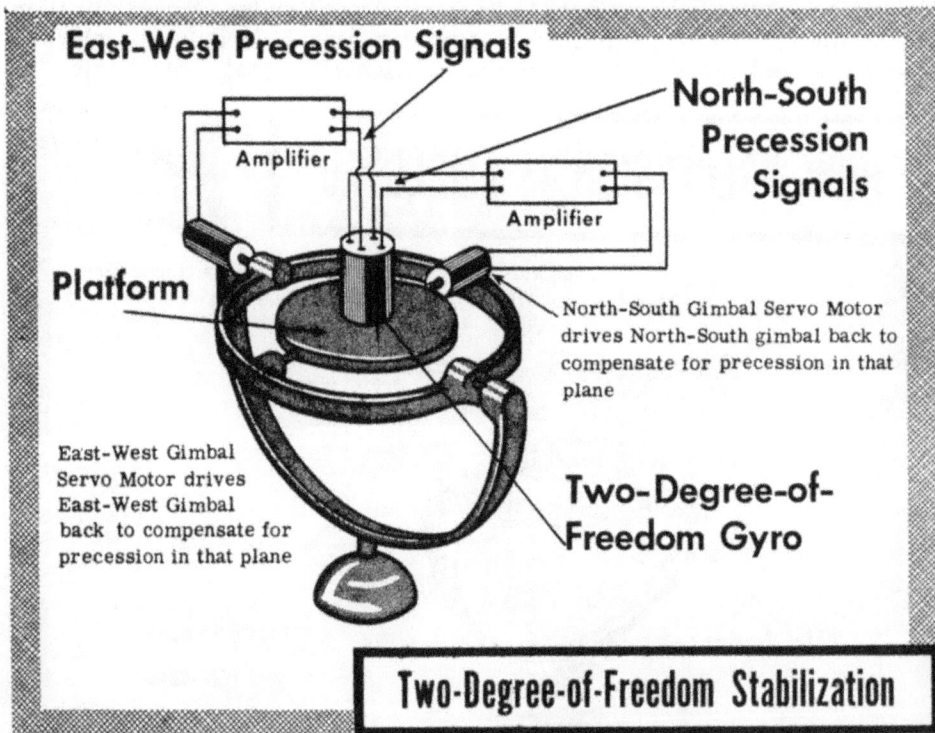

East-West Precession Signals

North-South Precession Signals

Amplifier

Amplifier

Platform

North-South Gimbal Servo Motor drives North-South gimbal back to compensate for precession in that plane

East-West Gimbal Servo Motor drives East-West Gimbal back to compensate for precession in that plane

Two-Degree-of-Freedom Gyro

Two-Degree-of-Freedom Stabilization

If we replaced the platform with a missile body, the principles would be the same as shown. However, several other elements would become necessary to provide control or guidance to the appropriate thrust or surface component. These might include valves, actuators, and a feedback loop.

The Navigational Possibilities of the Gyroscope

Although the gyroscope offered features which appeared to link it closely with navigational possibilities as related to motion about the surface of the Earth, it was not until the beginning of the twentieth century that any major applications were made. Three factors influenced this situation:

1. The need for a suitable means for spinning the rotor at high velocities. It is not uncommon in present-day gyroscopes to find rotor speeds in excess of 20,000 rpm.
2. The development of low-friction bearings or other means of supporting the rotor at these speeds as well as providing the lowest possible friction at the gimbal mounts.
3. The development of suitable arrangements for utilizing the gimbal motions for control purposes. Any appreciable restraining forces on the gimbal movements will introduce errors.

These three factors continue to dictate the direction of gyroscope development despite the many applications which have been realized to date. The continuing demand for smaller gyros with better performance characteristics keeps all of these problems alive as the designer proceeds from one application to another.

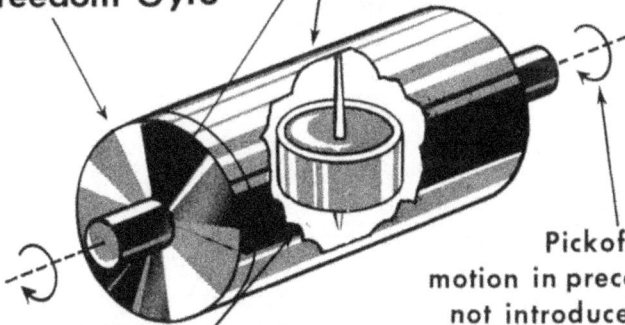

Fundamental Requirements for Gyroscope Usefulness

Gimbal Can of Single-Degree-of-Freedom Gyro

Bearing friction must be reduced to absolute minimum

Pickoff of gimbal motion in precession must not introduce disturbing torques

Gyro-Rotor must be driven at high speeds

Gyro Drift

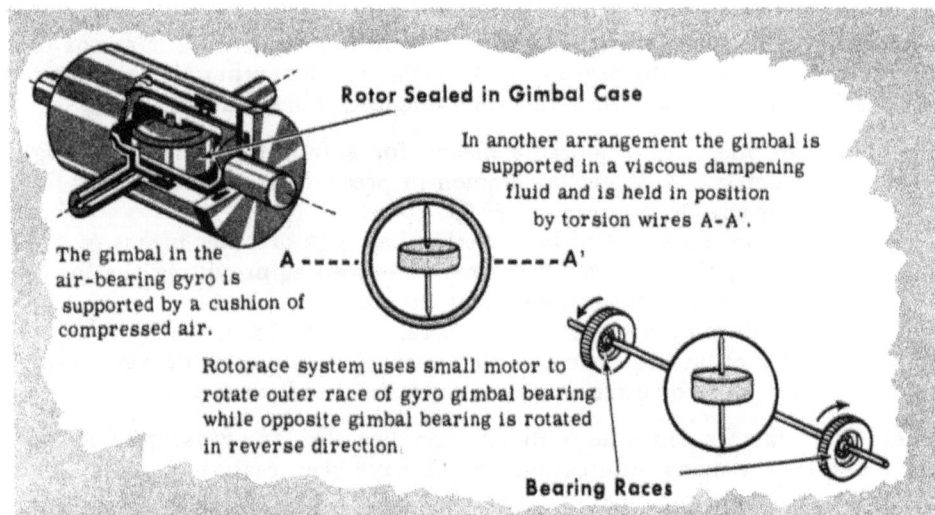

Rotor Sealed in Gimbal Case

In another arrangement the gimbal is supported in a viscous dampening fluid and is held in position by torsion wires A-A'.

The gimbal in the air-bearing gyro is supported by a cushion of compressed air.

A - - - - A'

Rotorace system uses small motor to rotate outer race of gyro gimbal bearing while opposite gimbal bearing is rotated in reverse direction.

Bearing Races

Although a perfect gyroscope would maintain true axial rigidity in space, it is apparent from the second and third factors mentioned above that perfection can only be approached but not realized as long as there is any friction in the rotor bearings or any unbalance from gimbal restraining forces. Hence, the continual quest for better performance in these areas. The degree to which the gyro axis deviates from true rigidity is called *drift*. If a gyro is to be used for aircraft stabilization over only a short period of time, a drift of 1° per hour might be quite acceptable, but if it is to be used as a basis for accurate navigation over a longer period of time or for greater accuracy, the drift must be reduced below .1° per hour. It is apparent that a gyro of a given drift-rate will perform best over a given range if the vehicle is moving at the maximum possible speed so that it will be in operation for the shortest possible time; better still, the problem is simplified if the gyro must be depended upon only during a relatively small part of the trip.

A variety of approaches are used to reduce bearing-friction and unbalance effects. Among these are air bearings where a cushion of air is used to support the parts away from metallic contact. The gyro rotor is also in some instances sealed in a gas-filled compartment to assure uniformity of environmental conditions. The gimbals may also be floated in a fluid and only supported by centering wires or rods. It is quite apparent that a high order of mechanical precision in parts fabrication is necessary to minimize drift effects due to unbalance in the parts related to the rotor and gimbal structures. Continual progress in gyro design and production engineering promise to produce the very low drift rates essential to long-range self-contained navigation.

GYROSCOPIC FUNDAMENTALS

Relating the Gyroscope to the Earth

Although it was recognized from the performance of the Foucault gyro that it had possibilities as a compass, it was not possible to produce a practical device until the development of high-speed electric motors and low-friction ball bearings. Herman Anschuetz Kaempfe of Germany produced the first practical gyro-compass in 1908 with Elmer Sperry of the U.S.A. following closely on his heels in 1911.

Actually the gyroscope is such a true space device in regard to the rigidity of its axis that many of the problems associated with its application on and about the Earth have been created by trying to relate it to this large rotating mass upon which man has lived and travelled during the many centuries of his existence. Specifically, some of the factors involved are:

1. The Earth's rotation about its North-South axis every 24 hours.
2. The relatively spherical shape of the Earth, which dictates travel and navigation along a curved path rather than in a straight line.
3. The Earth's gravitational field, which produces an acceleration of approximately 32 ft per sec^2 for a freely falling body in this field.

The GYROSCOPE as a NAVIGATIONAL DEVICE

GYRO POINTS TO A FIXED DIRECTION IN SPACE

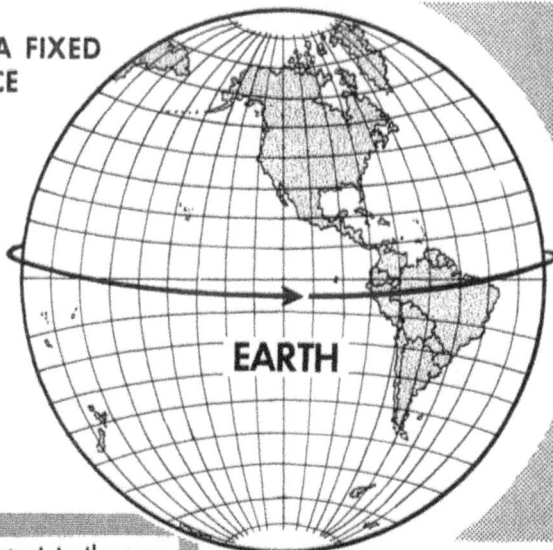

EARTH

The earth rotates with respect to the gyro.
Because the gyro is referenced to a point in space, it must be related to the earth to provide useful navigational data.

GYROSCOPIC FUNDAMENTALS

Gyroscopic Arrangements Related to the Earth

Two basic gyroscopic arrangements related to the surface of the Earth which have been widely applied are best described by considering the situation which prevails at one of the Poles and at the Equator. If a gyroscope is located at the North Pole, with its axis at right angles to the polar axis, the gyro axis will appear to make one complete revolution during a 24 hour period. However, in reality the gyro axis remains fixed in space, and the observer and the Earth rotate around it.

Gyro at North   Pole with Axis   at Right Angles   to Polar Axis

O HOURS   12 HOURS   24 HOURS

If the gyro axis at the North Pole is oriented into the same direction as the Earth's spin axis, there will be no apparent change in direction during a 24 hour period.

Gyro at North   Pole with Axis   Aligned to   Earth's Spin Axis

O HOURS   12HOURS   24HOURS

If we place a gyroscope at the Equator with its rotor axis horizontal and aligned East-West, the Earth's rotation every 24 hours will give the rotor axis an apparent vertical rotation through 360°. After 6 hours and 18 hours,

Gyro at Equator with Rotor Axis Pointing East-West

O HOURS   6 HOURS   18 HOURS

the rotor will be aligned as a vertical gyro. However, this apparent realignment is only relative to the Earth, since no axial deviation in space occurs.

Conversely, if we place a gyro at the Equator with its rotor axis horizontal and aligned North-South, no apparent tilt will be observed. Both its rotor axis and the Earth's spin axis are in the same plane.

Gyro at Equator with Rotor Axis Pointing North-South

NOON   MIDNIGHT   NOON

GYROSCOPIC FUNDAMENTALS

The Directional Gyroscope

It is apparent from these examples as well as from the required corrections for navigating any vehicle relative to the surface of the Earth that two principal gyro orientations would be applicable: (1) The *horizontal gyro* with its spin axis parallel to the surface of the Earth (2) The *vertical gyro* with its spin axis perpendicular to the Earth's surface. The former is known as the *directional gyro* and the latter as a *vertical gyro*.

This application of the directional gyros was ushered in with the development of the gyro-compass. In this application a correction is necessary to keep its axis in the proper direction to indicate the North-South direction. In slow-moving vehicles, such as ships, it is possible to achieve this correction to a degree by bottom-heavy rotors and cases or by a so-called *mercury ballistic* which causes the gyro to precess toward the Meridian.

However, in aircraft, where speeds are always significant, it is necessary to *slave* the gyro to a magnetic-compass transmitter. The gyro provides the necessary stability for short periods of flight time, and is continually monitored against the magnetic compass for long-period deviations. In a control function the directional gyro is a vital part of any autopilot system for sensing course deviations. It has also been used for stabilizing radar and bombing systems where its sensed information is applied through a suitable servo-loop.

Although the applications of directional gyros in aircraft have been numerous indeed, in missile applications the autopilot has been the principal one. In many missile applications it is necessary to utilize three gyros in a so-called *stable platform* and their axes may be oriented in directions other than horizontal. Stable platforms are also used in aircraft for various stabilization applications.

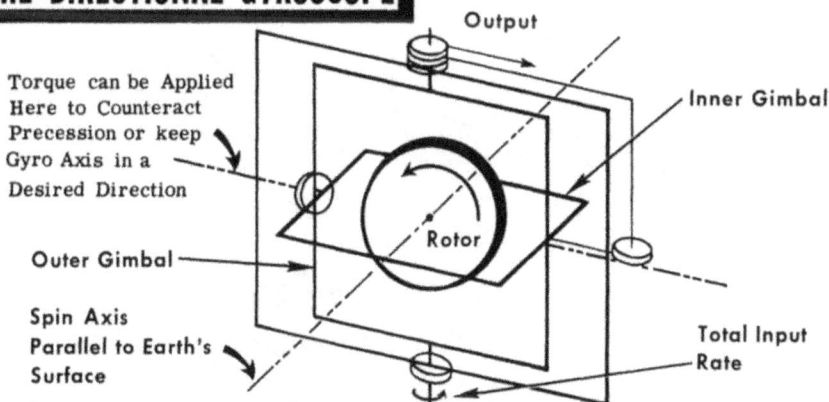

THE DIRECTIONAL GYROSCOPE

Output

Torque can be Applied Here to Counteract Precession or keep Gyro Axis in a Desired Direction

Inner Gimbal

Rotor

Outer Gimbal

Spin Axis Parallel to Earth's Surface

Total Input Rate

The Vertical Gyroscope

The vertical gyro has found numerous applications in aviation, marine, and ordnance fields, where position relative to a gravity reference is essential to precise and accurate operation of many indication, control, and stabilization systems. It is also employed as a control unit for numerous panel-mounted indicating instruments which visually display roll and pitch attitudes. Extensive use has also been made in radar-antenna platform stabilization.

In general, a pendulous reference is required to keep the rotor axle of this type of gyro vertical, regardless of the motion of the aircraft, the Earth's rotation, and the drift due to mechanical imperfections in the gyro. Vertical gyros are combined with directional gyros in autopilot systems to sense the necessary information for pitch and yaw stabilization relative to the desired flight path.

In missile applications, its principal use is in autopilots. The principal gyro functions are generally combined in stable platforms rather than being employed separately as directional or vertical gyros. Actually, both of these types as applied in aircraft are intended to provide stable operation only over relatively short periods of time without correction from either horizontal or vertical references. Hence, these gyros need not have exceedingly low-drift rates as would be required for highly accurate missile guidance applications.

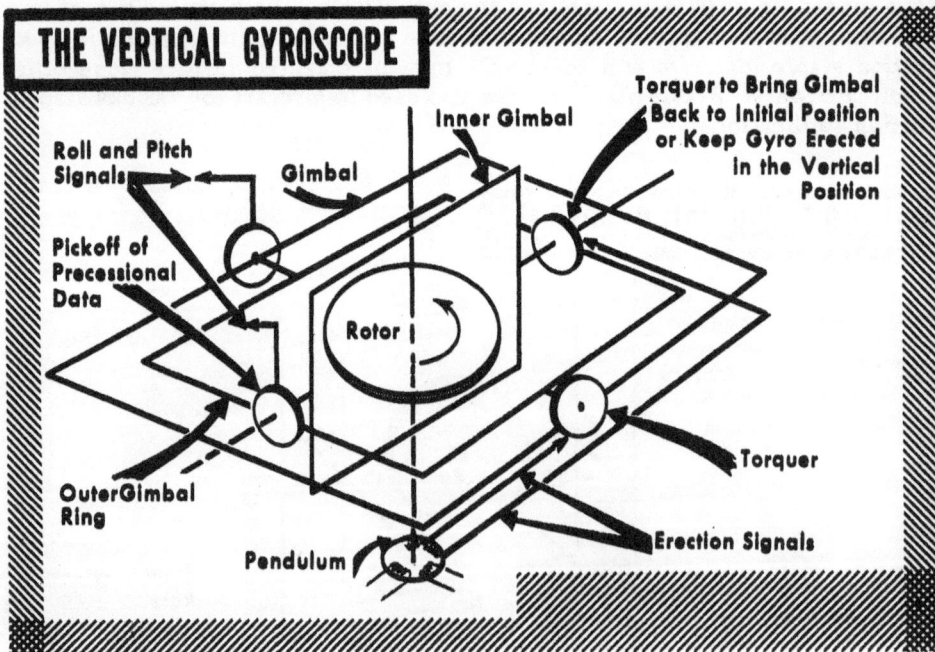

THE VERTICAL GYROSCOPE

Roll and Pitch Signals

Gimbal

Inner Gimbal

Torquer to Bring Gimbal Back to Initial Position or Keep Gyro Erected in the Vertical Position

Pickoff of Precessional Data

Rotor

Torquer

Outer Gimbal Ring

Pendulum

Erection Signals

The Free Gyroscope

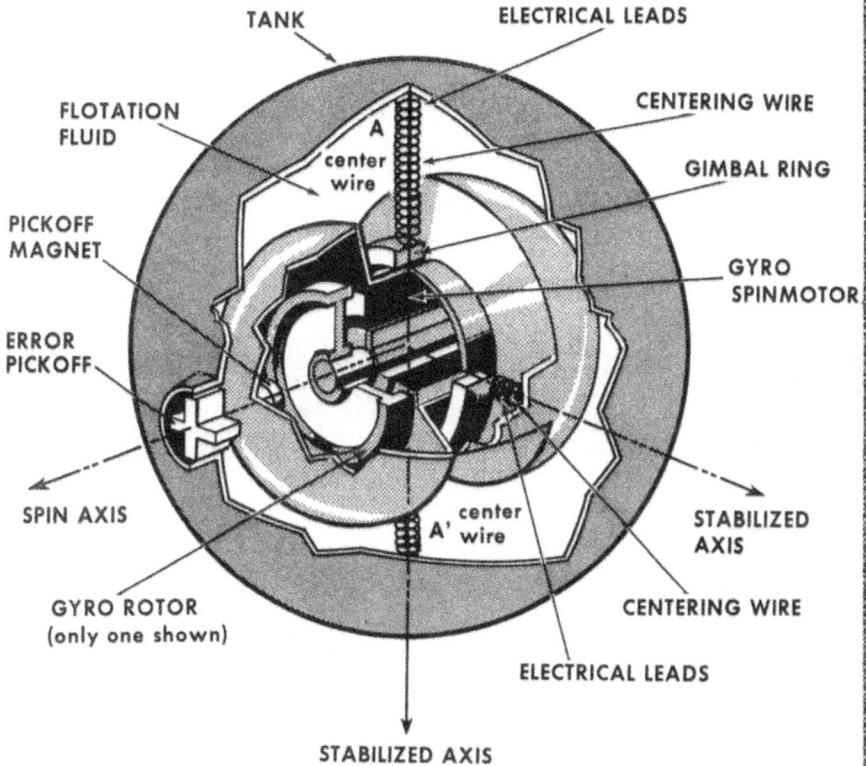

A TYPICAL FREE GYROSCOPE

TANK

ELECTRICAL LEADS

FLOTATION FLUID

CENTERING WIRE

A center wire

GIMBAL RING

PICKOFF MAGNET

GYRO SPINMOTOR

ERROR PICKOFF

SPIN AXIS

A' center wire

STABILIZED AXIS

GYRO ROTOR (only one shown)

CENTERING WIRE

ELECTRICAL LEADS

STABILIZED AXIS

A free gyro differs from a directional or vertical gyro primarily in the fact that its axis may be oriented in any direction relative to the surface of the Earth. The rotor is caged or held in the desired direction, then uncaged or released to give the necessary gyro-action for stabilization. A typical application is for roll-stabilization. However, the free gyroscope finds a wide range of applications. For instance, the two-degree-of–freedom gyros used for missile-stabilization in a typical stable platform are free gyros.

(1-85)

GYROSCOPIC FUNDAMENTALS

The Rate Gyroscope

Although directional, vertical, and free gyros can sense the information necessary to determine a position in space, it is particularly important, in stabilizing any vehicle, to know the rate at which deviations and/or corrections are being applied as in the case of the autopilot. This parameter may be determined by utilizing rate gyros, which can be made by applying a restraining spring to the moving gimbal of a single-degree-of-freedom gyro. When the torque developed by turning forces is applied to the gyro, it will precess. However, the restraining springs will allow the gimbal to move only through a small arc proportional to the angular rate along the input axis.

Two types of rate gyros are in extensive use. One type provides a voltage output proportional to the input rate and the gimbal displacement. The other type acts as a switch which closes a circuit when the rate reaches a predetermined value. Both units may be mounted in any position which correctly orients the input axis.

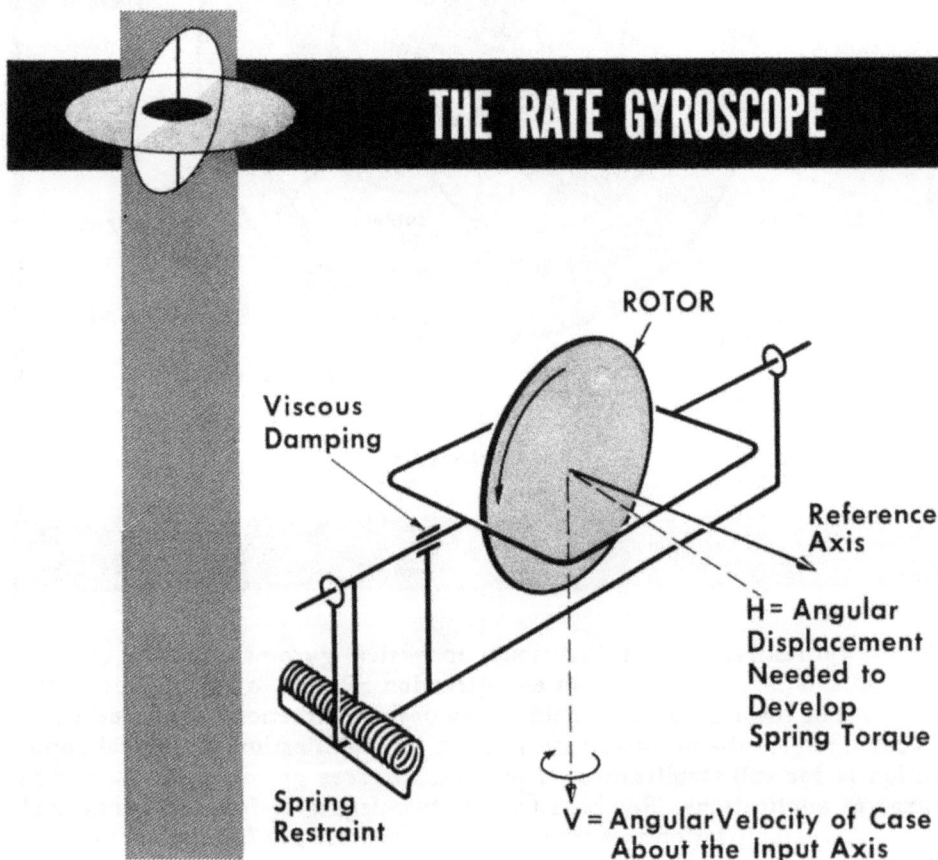

THE RATE GYROSCOPE

ROTOR

Viscous Damping

Reference Axis

H= Angular Displacement Needed to Develop Spring Torque

Spring Restraint

V = AngularVelocity of Case About the Input Axis

The Rate Gyroscope (contd.)

Another type of rate gyro is the *floated integrating* type. It is a single-degree-of-freedom unit, in which there is no intentional restraint except viscous friction from a fluid in which the gimbal is immersed. It is also referred to as a torque-summing computer which converts input turning rates into torques that are time-integrated into gimbal displacements through a viscous fluid. Information proportional to angular displacement of the float relative to the case is obtained from a pickoff on the precession axis. The pickoff signal is generated in stator windings inside of a movable armature attached at one end of the gimbal. In some designs, the armature may rotate inside the stator.

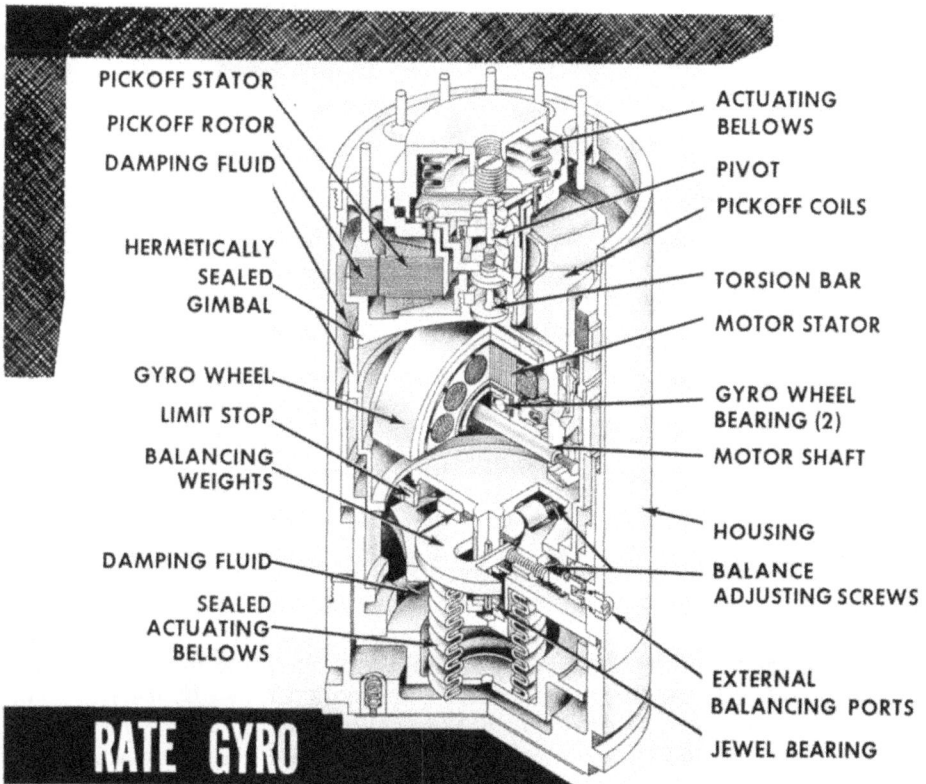

RATE GYRO

Rate measuring gyros are widely used to provide voltage outputs proportional to the rate of turn in the following applications: airborne autopilots, fire control systems, oscillation dampers, missile stabilization and telemetering systems. Gyro rate switches are used: (a) to cut out erection in vertical gyros and stable platforms during turns, or (b) to cut out slaving voltages in directional gyros during turns.

The Floated Integrating Gyros

Floated integrating (IG) gyros are widely used as basic sensors of angular rate and angular position in a number of aircraft and missile applications — such as, in reference stabilization and navigational platforms for flight control systems, stabilization and tracking for airborne and shipborne fire control systems and rate measurement of platform deviation in missile guidance.

Floated, Single-Degree-of-Freedom, Integrating Gyro

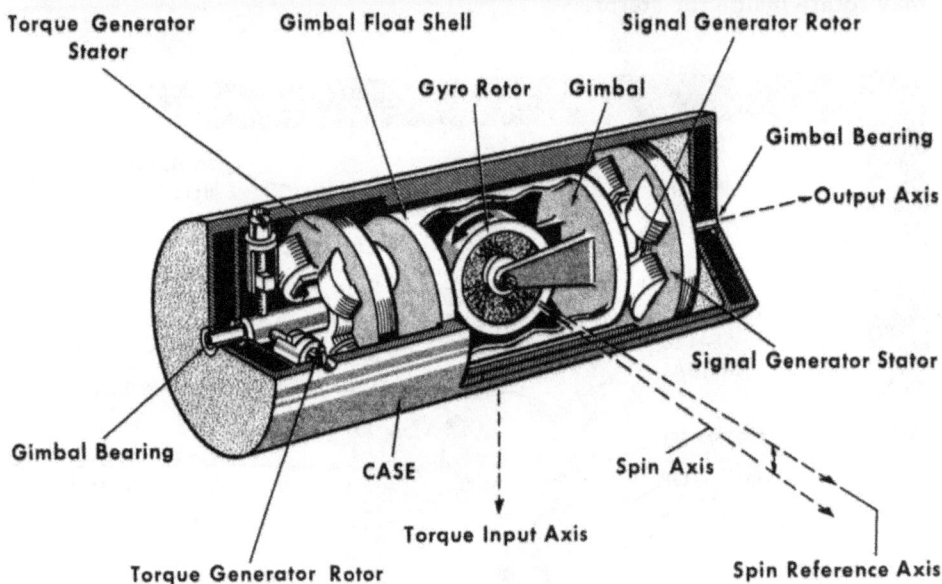

A typical floated integrating (IG) gyro has its rotor in a sealed chamber filled with a gas — such as helium. The sealed chamber which really forms the single axis gimbal is mounted in a viscous fluid — such as fluorolube. The signal generator develops output voltages proportional to the precession of the gimbal, and the torque generator provides a means of inserting command signals to restore the gimbal to the proper position to balance precessional forces. In other constructions both the torque generator and signal generator are located at one end of the gimbal axis.

This type of gyro often utilizes a fluid which must be heated before it is placed in operation, due to the fact that the most suitable fluids solidify at room temperatures. Thus, a warm-up period is required before any system with which they are associated can become operational. Certain types of integrating gyros are designated (HIG) gyros for hermetically sealed integrating gyros.

Stable Platforms

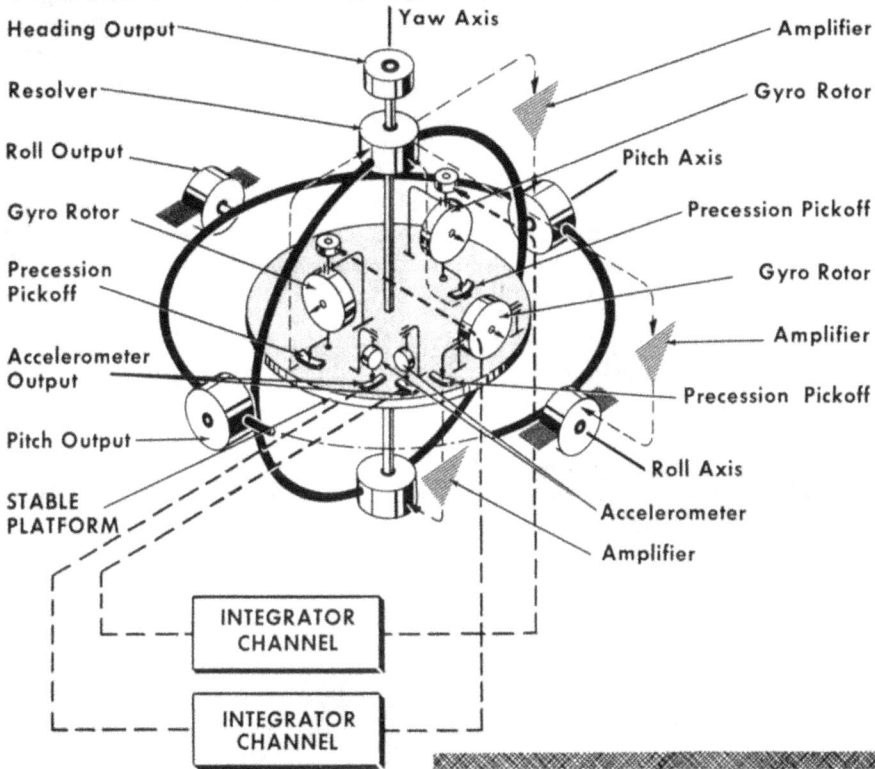

THE THREE-GYRO STABLE PLATFORM

Heading Output — Yaw Axis — Amplifier

Resolver — Gyro Rotor

Roll Output — Pitch Axis

Gyro Rotor — Precession Pickoff

Precession Pickoff — Gyro Rotor

Accelerometer Output — Amplifier

Pitch Output — Precession Pickoff

STABLE PLATFORM — Roll Axis

Accelerometer

Amplifier

INTEGRATOR CHANNEL

INTEGRATOR CHANNEL

In recent years it has become more popular to utilize so-called stable platforms — consisting of either three single-degree-of-freedom gyros or a pair of two-degree-of-freedom gyros — to stabilize a system, rather than a large number of separate directional, vertical, and rate gyros to stabilize the individual units of a system. Functionally, a stable platform performs as a vertical and directional gyro combined, with a free gyro thrown in for good measure. However, it separates the gyro elements from the frame of the aircraft or missile into an integral package.

This basic design refinement improves the overall accuracy of a stable platform system to a higher degree than a system composed of a vertical and directional gyro mounted to the body of the controlled craft. The gimbaling error found in a conventional gyro is eliminated, and drift rates are lower than in conventional gyros because sensitive axis bearings are operated over only a limited range. Increased accuracy combined with the advantage of a low-volume single package make this approach a desirable one.

Stable Platforms (contd.)

The attitude information obtained from the gimbal axis transmitters can be used in several applications. Among these are autopilot reference signals, antenna stabilization, heading reference indication and verticality indication. As will be shown in following pages, the stable platform is important to inertial navigation as an accurately positioned mounting for precision-accelerometers. A three-gyro, three-gimbal platform has 360° of freedom about the roll and azimuth axes, but is limited to ±85° in pitch. By increasing the number of gimbals to four, 360° of freedom about all axes may be realized. Thus, with an additional gimbal the stable platform becomes very applicable to high performance, all attitude aircraft. However, for ballistic and cruise type missiles, three-gimbal platforms can be quite effective because they do not maneuver widely in all attitudes of flight.

For long-range navigation along the surface of the earth, such as for ocean-going vessels, it is considered necessary to employ a five-gimbal platform, to provide for longitude and latitude corrections.

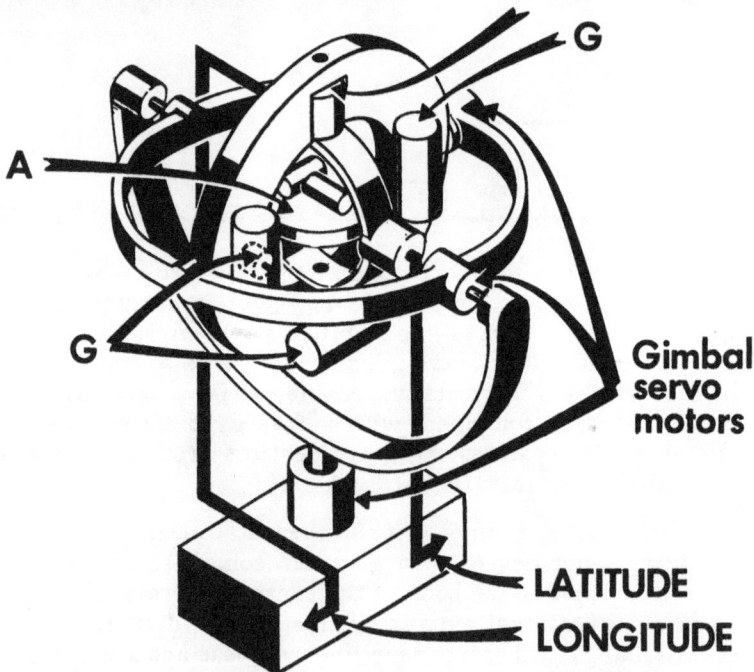

Gravity-inertia system

Longitude and Earth's rotation motor

Gimbal servo motors

LATITUDE
LONGITUDE

G-- Single degree of freedom gyroscope
A-- Accelerometer package

Introduction to Inertial Guidance

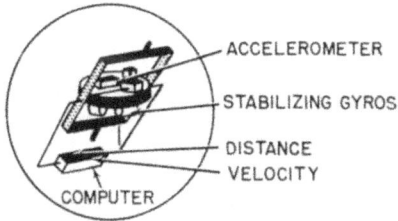

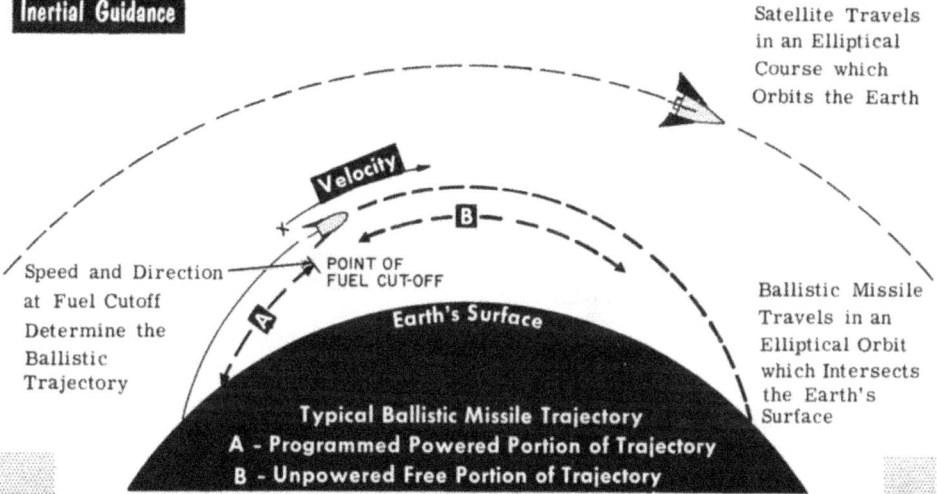

ACCELEROMETER

STABILIZING GYROS

DISTANCE

VELOCITY

COMPUTER

APPLICATIONS OF INERTIAL GUIDANCE

Inertial Guidance

Satellite Travels in an Elliptical Course which Orbits the Earth

Velocity

B

POINT OF FUEL CUT-OFF

Speed and Direction at Fuel Cutoff Determine the Ballistic Trajectory

A

Earth's Surface

Ballistic Missile Travels in an Elliptical Orbit which Intersects the Earth's Surface

Typical Ballistic Missile Trajectory
A - Programmed Powered Portion of Trajectory
B - Unpowered Free Portion of Trajectory

The range of a rocket in a typical ballistic path is to first approximation a function of its velocity (speed and direction) at the fuel cutoff or all-burnt point. Suppose we think of the ballistic trajectory as an elliptical orbit about the center of the Earth, which intersects the surface of the Earth, rather than a nonintersecting satellite orbit. Since velocity is a determining factor in the satellite orbit, it also determines the ballistic ellipse and its intersecting point with the Earth's surface.

Previously we discussed how a guided missile could be brought to a point in space by a radar guide-beam technique, and how the point of fuel cutoff could be determined by measuring the velocity with a Doppler-radar system.

The initial application of another technique of a completely self-contained nature within the missile was made in one form of V-2 guidance. The missile contained a gyro stabilizer that kept it in the proper yaw and roll positions. The pitch gyro was programmed to tilt the rocket into a ballistic trajectory. To establish the proper point for fuel cutoff an accelerometer was used to determine the velocity at which this cutoff should occur.

Accelerometer Fundamentals

One of the most elementary forms of an accelerometer consists of a weight suspended between two springs. Such a simple device is never employed in a practical system. If a force is applied to move the accelerometer to the right the mass **M** will tend to remain at rest, will compress the spring portion at the left of **M**, and stretch it to the right of **M**. If a pointer is attached to **M**, it will indicate a measure of the acceleration on the scale as it moves to the left. If the applied force was in the opposite direction, the mass **M** would move in the opposite direction. Thus, the amount and direction of the acceleration can be obtained. The product of acceleration and time equals velocity. If we have a means of measuring acceleration and time, hence a means of obtaining their product by multiplication, we can obtain the velocity of a vehicle, rocket, or missile with which these measuring devices are associated.

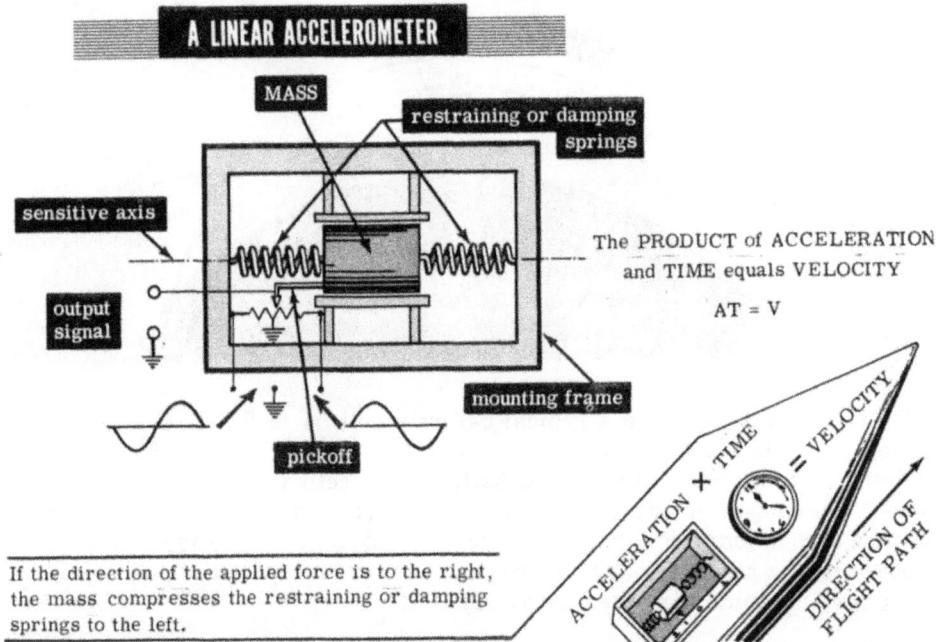

A LINEAR ACCELEROMETER

MASS

restraining or damping springs

sensitive axis

output signal

pickoff

mounting frame

The PRODUCT of ACCELERATION and TIME equals VELOCITY

AT = V

ACCELERATION × TIME = VELOCITY

DIRECTION OF FLIGHT PATH

If the direction of the applied force is to the right, the mass compresses the restraining or damping springs to the left.

An accelerometer, placed along the flight axis of a missile together with a clock and computer, can measure the missile's velocity in that direction. By comparing the actual velocity to the predetermined value required to provide the desired range of flight, the fuel can be cut off at the right point. This example is one of the simplest forms of inertial guidance. It is generally employed in a somewhat more complex form, but the same fundamental principles apply. The velocity is obtained by summing or integrating the accelerations over a period of time.

The Unbalanced Gyro Accelerometer

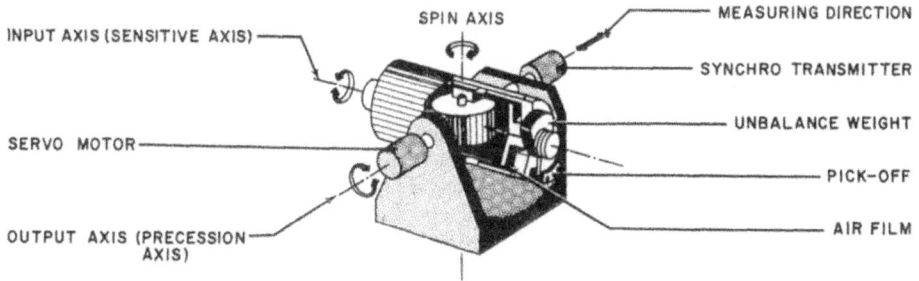

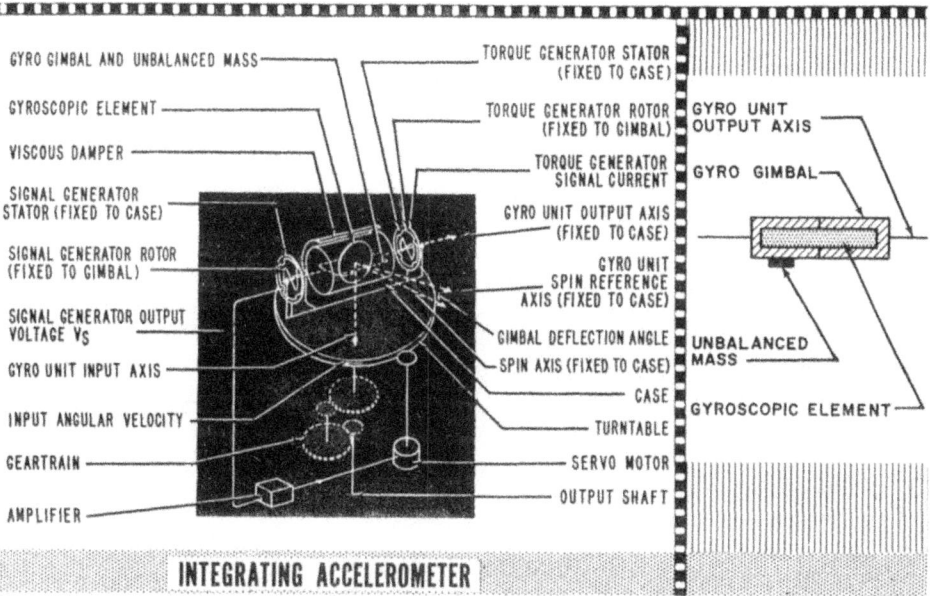

UNBALANCED GYRO ACCELEROMETER

INPUT AXIS (SENSITIVE AXIS)

SPIN AXIS

MEASURING DIRECTION

SYNCHRO TRANSMITTER

UNBALANCE WEIGHT

SERVO MOTOR

PICK-OFF

OUTPUT AXIS (PRECESSION AXIS)

AIR FILM

GYRO GIMBAL AND UNBALANCED MASS

GYROSCOPIC ELEMENT

VISCOUS DAMPER

SIGNAL GENERATOR STATOR (FIXED TO CASE)

SIGNAL GENERATOR ROTOR (FIXED TO GIMBAL)

SIGNAL GENERATOR OUTPUT VOLTAGE V_S

GYRO UNIT INPUT AXIS

INPUT ANGULAR VELOCITY

GEARTRAIN

AMPLIFIER

TORQUE GENERATOR STATOR (FIXED TO CASE)

TORQUE GENERATOR ROTOR (FIXED TO GIMBAL)

TORQUE GENERATOR SIGNAL CURRENT

GYRO UNIT OUTPUT AXIS (FIXED TO CASE)

GYRO UNIT SPIN REFERENCE AXIS (FIXED TO CASE)

GIMBAL DEFLECTION ANGLE

SPIN AXIS (FIXED TO CASE)

CASE

TURNTABLE

SERVO MOTOR

OUTPUT SHAFT

GYRO UNIT OUTPUT AXIS

GYRO GIMBAL

UNBALANCED MASS

GYROSCOPIC ELEMENT

INTEGRATING ACCELEROMETER

Before giving further consideration to accelerometer applications in inertial navigation systems, let us look at some of the forms which a guidance accelerometer may take. In some of the V-2 development work, both gyroscopes and pendulums were employed. In the gyroscopic accelerometer use was made of the fact that an unbalanced gyro, under the influence of an acceleration, precesses at a rate proportional to this acceleration. The angle of precession is proportional to the integral or summation of all accelerations over a period of time. The fuel cutoff was actuated when the gyro reached a predetermined angle of precession, and when it closed a properly adjusted contact.

The Pendulum and Other Types of Accelerometers

In one pendulum-type accelerometer a swinging mass is suspended in an electromagnetic field. When the pendulum moves from its vertical position under the influence of an acceleration, a bridge circuit is unbalanced. Thus, a current is caused to flow in the moving coil which restricts the pendulum motion. The condition of equilibrium is such that the torque caused by the acceleration and the torque of the moving coil balance each other. The moving coil current is also proportional to the acceleration. This current is integrated by means of an electrolytic cell specially developed for this purpose. The charge of the cell (B) during the flight is proportional to the velocity. Before launching, the cell was charged with another voltage (A), corresponding to the desired fuel cutoff velocity, in an opposite sense, so that fuel cutoff is obtained when A-B passes through zero.

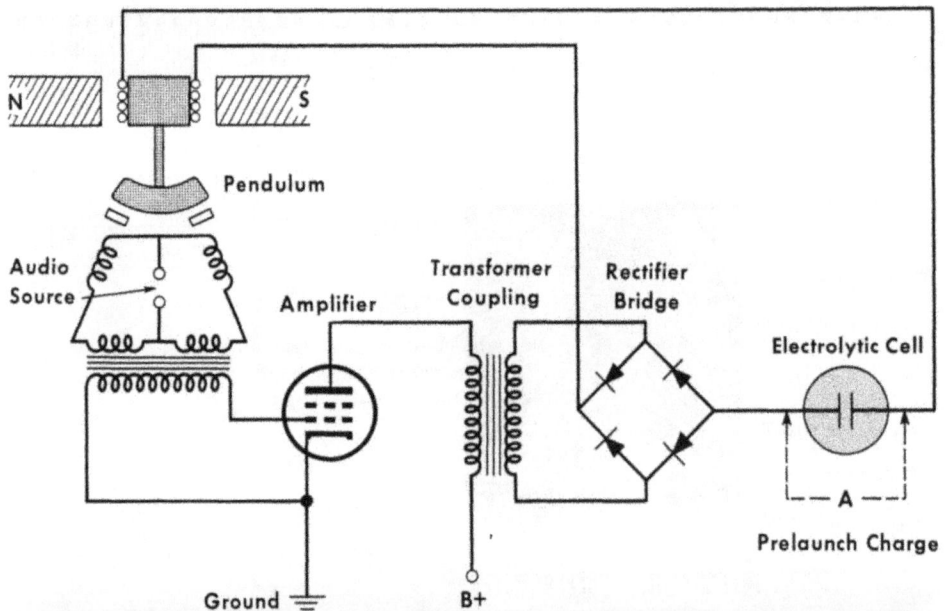

The Pendulum – Type Accelerometer

Many other types of accelerometers have since been proposed and undoubtedly developed to some degree of perfection. These include piezoelectric, strain gage, vibrating wire, variable inductance, variable capacitance, electronic element transducer, differential transformer and potentiometer types. For precise inertial guidance and navigation applications, accelerometers must respond to very minute changes of acceleration.

Accelerometer Applications to Guidance

In their first applications to missile guidance, accelerometers were fixed to the missile body and were only intended to provide velocity data in the direction of the flight path. Such an arrangement did not permit the use of an accelerometer in the lateral plane for guidance. To remedy this situation the accelerometers were mounted on a stabilized platform. By stabilizing such a platform with three gyros, effective along the yaw, roll and pitch axes, its direction in space could be controlled. With this arrangement it is possible to utilize the accelerometer along longitudinal and lateral axes to determine either velocity by a single integration or the distance travelled by two successive integrations in these directions.

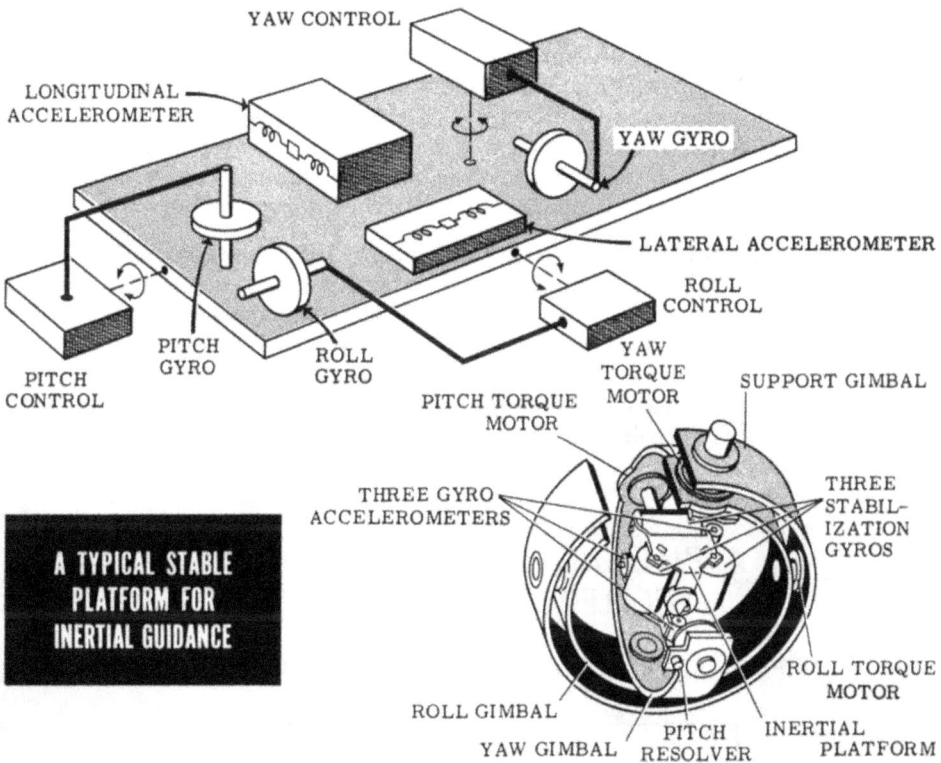

A TYPICAL STABLE PLATFORM FOR INERTIAL GUIDANCE

It is possible to use these outputs to correct for lateral or longitudinal deviations from a desired flight path, and apply corrective signals through appropriate servo systems to keep a missile in a desired trajectory. In the case of the precessing gyro accelerometer-velocity data could be obtained directly for application to an analog computer. In the case of a vibrating-string accelerometer the output appears in the form of a frequency, and the first integration is performed by counting in a digital computer.

Accelerometer Applications to Guidance (contd.)

Up to this point we have given a rather simplified explanation of the inertial navigation problem. However, due to the Earth's gravity field accelerometers do not indicate the true acceleration relative to the Earth. This situation results from the fact that gravity exerts an equal force upon the accelerometer mass as well as upon its container. To compensate for this behavior it is necessary to compute the acceleration of gravity continuously, and add it to the output of the accelerometers. From this we can obtain the true resultant accelerations which can then be doubly integrated to obtain position.

To measure the nongravitational acceleration, one can employ three linear accelerometers with mutually perpendicularly oriented axes. This arrangement will just suffice to determine completely the nongravitational accelerations. Since three accelerometers must be stabilized, we choose a nonrotating inertial frame of reference which maintains a fixed inertial attitude, when perfectly balanced, to eliminate disturbing torques. Since the gravity-vector term depends upon position, the output of the integrators is fed to a gravity computer which determines the gravity component at each position. The gravity term is then combined with the accelerometer data.

The output after the two successive integrations supplies the position in the inertial frame of reference. Since we are interested in position with respect to a rotating Earth, the output is fed into a transformation computer which transforms the output to position with respect to the Earth taking into account the rotation of the Earth (Coriolis acceleration). In the case of a ballistic missile the inertial computer must be combined with a ballistic computer which controls the engine thrust to provide fuel cutoff at the correct velocity.

APPLICATIONS OF ACCELEROMETER OUTPUTS

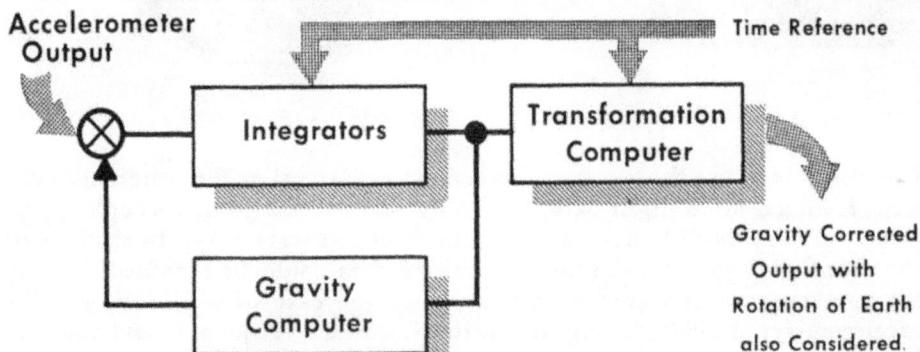

Accelerometer Output

Time Reference

Integrators

Transformation Computer

Gravity Computer

Gravity Corrected Output with Rotation of Earth also Considered.

Example of a True-Inertia System for Ballistic Launchings

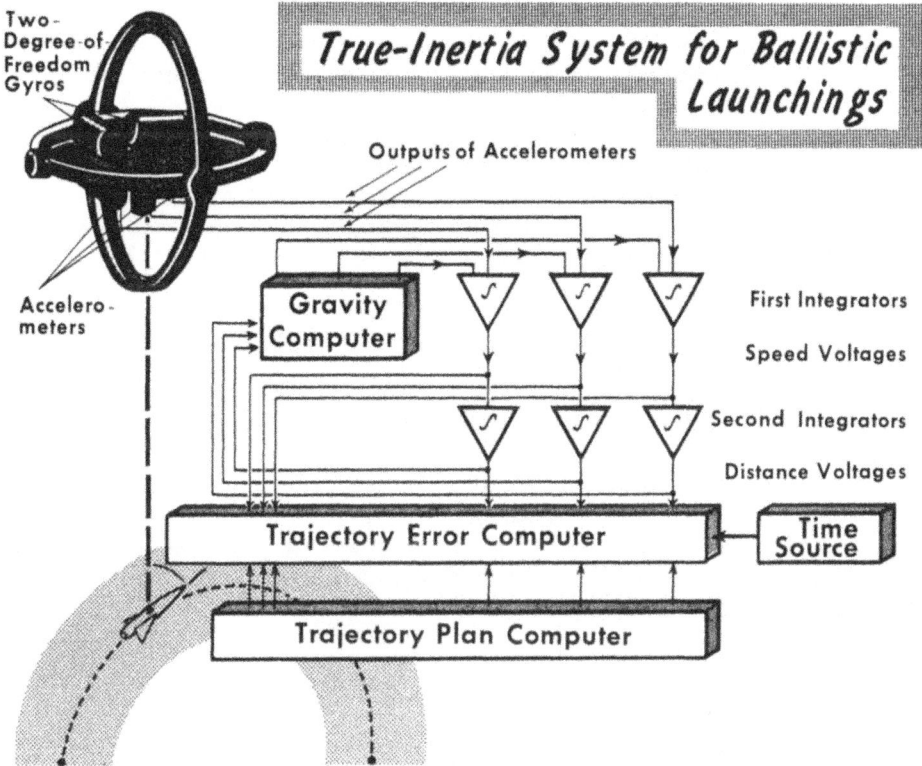

In the application of a true-inertia system to a V-2 or any other ballistic missile, the platform and the missile should be oriented at the launch point so that one platform-axis is normal to the trajectory path, and the other two axes are vertical and horizontal, respectively, in the trajectory plane. Except for gyro drift, the platform will maintain its initial orientation during flight. Any movement of the missile out of the trajectory plane caused by winds or thrust misalignment will result in a signal from the appropriate accelerometer to return it to the correct path. Due to the gravity component of the acceleration corrections are fed into the system for this factor.

Another function of the appropriate accelerometer is to provide data for cutting off the rocket fuel at the proper point in the trajectory. During the powered portion of the missile's trajectory comparisons should be made between its actual trajectory and the planned path in the trajectory error computer, and thrust can be directed to correct toward the planned trajectory. Calculations prior to launching would have provided the basis for the planned ballistic path to intersect the target. Inertial guidance systems for present-day missiles are based on these principles.

Precessed-Gravity Inertia Systems

In dealing with inertial guidance systems for the longer ranges of cruise-type crafts about the circumference of the Earth, we immediately encounter some of the same problems that we had with gyroscopes. Both the single gyro and the stabilized platform are really referenced to a point in space because of the rigidity of the gyroscopic axis. So, in traversing the Earth's surface, it is necessary to employ a means for keeping the platform tangent to the Earth's surface or perpendicular to a line drawn from the platform to the center of the Earth. In order to realize this situation it is necessary to depress the forward end of the platform at a rate proportional to the velocity of the missile or aircraft around the Earth.

In an inertial system the platform is kept level by dividing in the computer the measured missile velocity by its distance from the center of the Earth (this distance is its altitude plus the Earth's radius). Such a factor can be expressed as the angular velocity of the missile in radians. If the pitch angle of the platform is changed at this same angular velocity, the platform will remain tangent to the Earth about the pitch axis. To accomplish this result, the pitch gyro is precessed according to the angular velocity of the missile about the Earth. Since the tilting of the platform provides an inherent compensation for gravity effects, this system is referred to as a "precessed-gravity inertial" type. A separate gravity computer is not required.

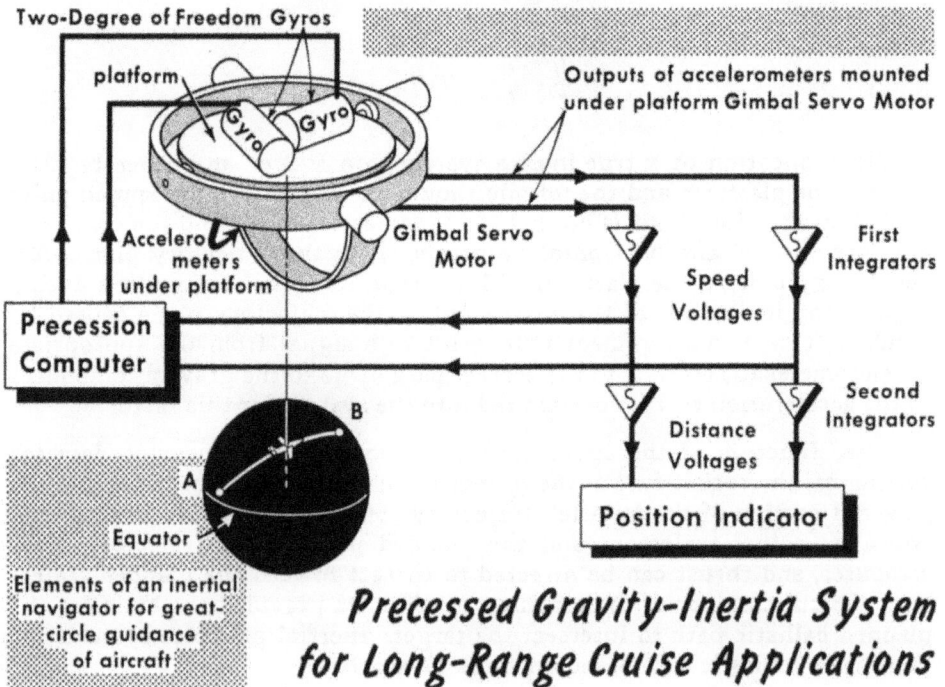

Precessed Gravity-Inertia System for Long-Range Cruise Applications

INERTIAL AND CELESTIAL NAVIGATION

Inertial Navigation Principles

Since the platform is tilted in proportion to the missile's velocity and is always kept at right angles to the Earth's surface, the angle of tilt is actually a measure of the distance traveled. Hence, a factor for determining the missile's position relative to its launching point is established, provided the missile is stabilized in yaw and roll. Therefore it is sometimes stated that inertial navigation really involves the measurement of the angle through which the platform must be rotated to keep it tangential with the Earth's surface and at right angles to the local vertical.

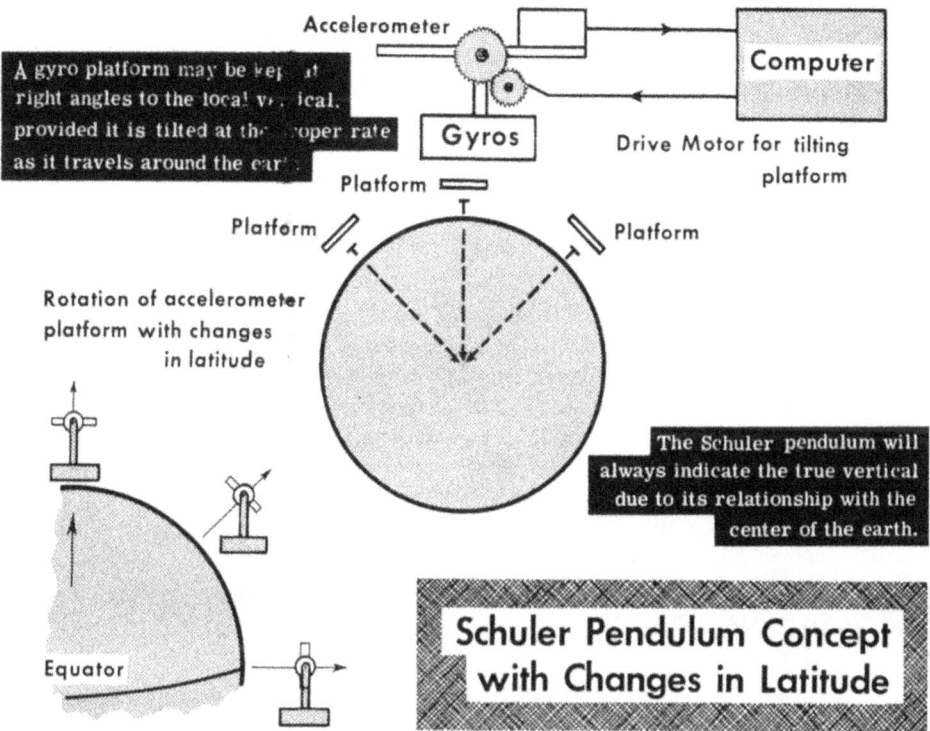

Accelerometer

Computer

A gyro platform may be kept at right angles to the local vertical, provided it is tilted at the proper rate as it travels around the earth.

Gyros

Drive Motor for tilting platform

Platform

Platform

Platform

Rotation of accelerometer platform with changes in latitude

The Schuler pendulum will always indicate the true vertical due to its relationship with the center of the earth.

Equator

Schuler Pendulum Concept with Changes in Latitude

In explaining inertial navigation principles, reference is often made to the *Schuler* pendulum — a hypothetical pendulum with a length equal to the radius of the Earth. If such a pendulum could be constructed, and if it were suspended from an object moving about the surface of the Earth, it would always point to the center of the Earth and indicate the local vertical. Since the inertial navigation platform must establish itself in a position relative to the local vertical, to measure the angular velocity of the missile, it is said to simulate a Schuler pendulum. The concept of this pendulum was first expressed in 1923 by Max Schuler, a German gyro expert.

Inertial Navigation Principles (contd.)

One of the principal problems in inertial navigation is the drift of the platform-stabilizing gyroscopes from their initial setting in space. Such a drift is produced primarily by bearing friction. Naturally this is a subject of considerable study and work toward improvement in the quest for a highly accurate self-contained system of navigation. It has been shown that a .01° per hour drift rate can result in an equal navigational error in degrees per hour. Therefore, somewhat lower drift rates are sought for the platform-stabilizing gyros of cruise-type vehicles covering long ranges.

The problem is less severe for ballistic missiles where the guidance is effective for only a few minutes. The drift may be compensated to a certain extent by circuits which feed back both fast and slow deviations to the platform for stabilization. The *slow loop* does not respond to any fast deviation, but sums the error during the complete flight. The *fast loop* responds to any rapid deviations in the platform. The slow loop sums the precession required to correct for the local vertical as well as the gyro drift. This feedback minimizes the gyro-drift effect and gives an oscillatory effect comparable to the period of a Schuler pendulum (84 minutes). This effect causes the platform to oscillate slightly about its position, cancelling the gyro drift to a certain extent rather than allowing it to accumulate in one direction.

Other factors of significance in inertial systems are resolution, linearity, and scale-factor accuracy of the accelerometers. Changes in temperature, vibration, effective gravity acceleration, and the position of components are also significant. Despite these difficulties inertial navigation offers complete automaticity, invulnerability to detection and jamming, and independence from ground-based aids. However, if external references can be tolerated for realignment of the system during extended flight periods the accuracy can be greatly improved.

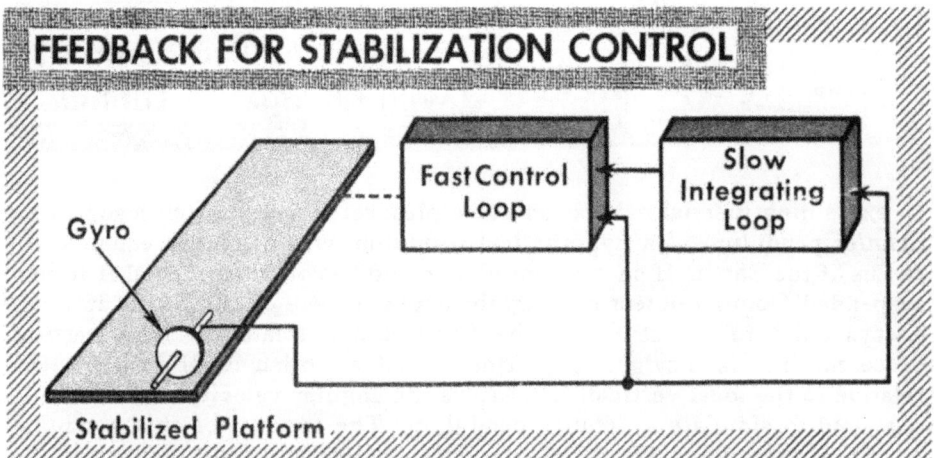

FEEDBACK FOR STABILIZATION CONTROL

Gyro

Fast Control Loop

Slow Integrating Loop

Stabilized Platform

INERTIAL AND CELESTIAL NAVIGATION

External Aids to Inertial Guidance and Navigation

The possibility of supplementing these systems with external data offers means for correcting gyro errors. With such an arrangement the inertial system can be utilized all of the time during the guidance or navigation phase, but at periodic intervals it can be supplemented with velocity or position data from radar or celestial sources depending upon the application.

In ballistic guidance systems the radar-inertial type is of particular interest. In such a system the inertial components need not have the highest accuracy capabilities, but can be corrected by radar-tracking data.

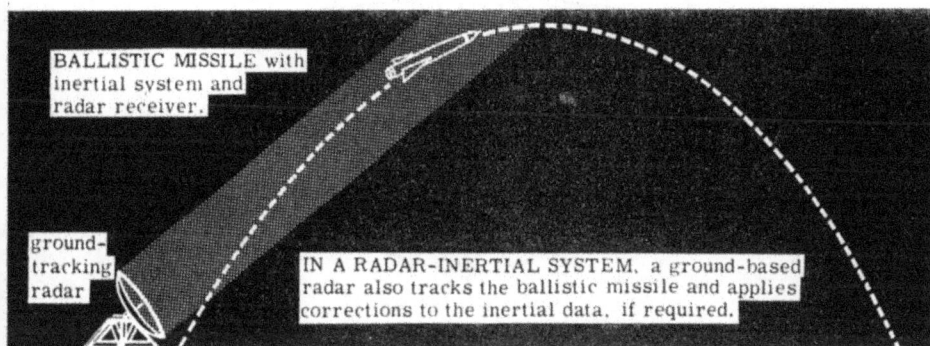

BALLISTIC MISSILE with inertial system and radar receiver.

ground-tracking radar

IN A RADAR-INERTIAL SYSTEM, a ground-based radar also tracks the ballistic missile and applies corrections to the inertial data, if required.

In long-range navigation applications two types of external references may be employed to assist inertial systems — namely, Doppler-radar and celestial tracking. In the Doppler-inertial system the radar equipment provides an additional source for velocity data so that comparisons can be made between

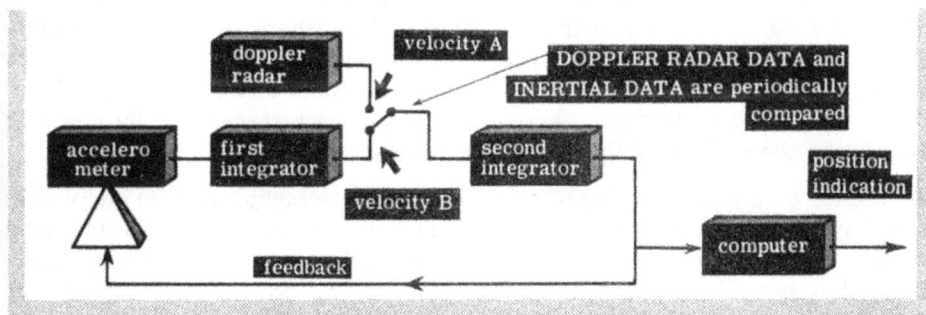

doppler radar — velocity A — DOPPLER RADAR DATA and INERTIAL DATA are periodically compared

accelerometer — first integrator — velocity B — second integrator — position indication — feedback — computer

the velocity measured by the pure inertial system and that obtained by the radar technique. In the celestial-inertial system the position of the vehicle is determined by an automatic star-tracker and fed in for comparison with the double integrated data of the inertial system.

Celestial Navigation

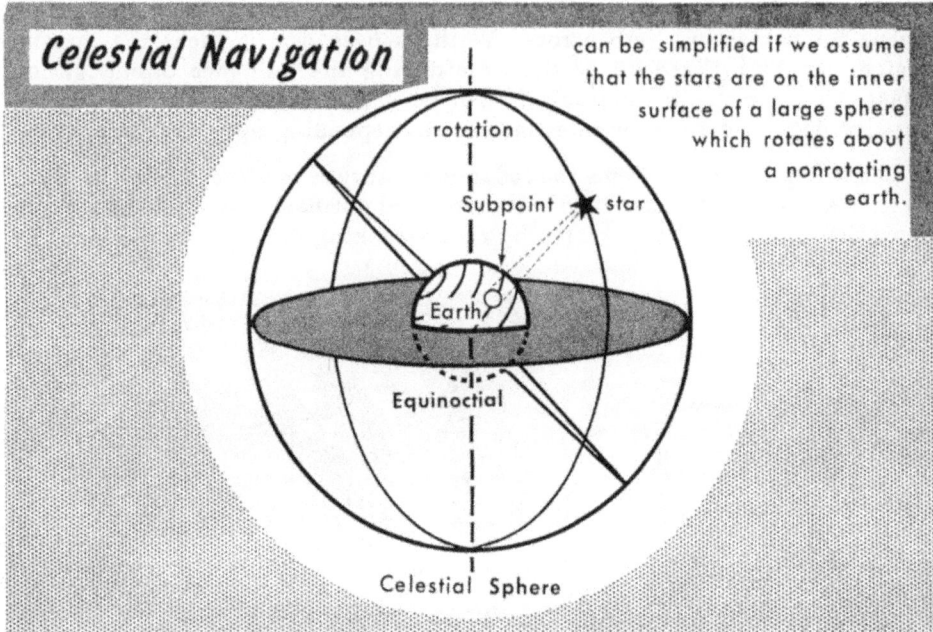

Celestial Navigation can be simplified if we assume that the stars are on the inner surface of a large sphere which rotates about a nonrotating earth.

rotation

Subpoint → ★ star

Earth

Equinoctial

Celestial Sphere

The concepts of celestial navigation can be simplified if we assume that the stars are on the inner surface of an infinitely large sphere which rotates about a nonrotating Earth. All of the stars appear to rotate westward together; however, the planets of the solar system appear to move in eccentric orbits. At any instant of time the stars on the celestial sphere appear over the Earth at corresponding subpoints. But their relative ground speeds are not the same; those nearest the Equator have the fastest apparent velocities of about 1000 mph; those in the Northern and Southern hemispheres have a slower apparent speed. In the zones of the major cities of the World, which lie between 60° North and 40° South, the minimum apparent star speed can be as low as about 500 mph. Actually it is the path traced out by any star relative to the surface of the Earth which becomes the tracking reference.

A missile or aircraft equipped with a suitable telescope and an automatic servo system to control its guiding surfaces as well as its fuel consumption could travel with a high degree of accuracy relative to such a path. However, it is not likely that a star path intersecting both the launching site and the target will be found. Therefore, rather than depending on one star path, it is better to track two stars — one whose path lies tangent to the course of the missile, the other whose path lies normal to the desired missile course. The information from the first star would be used for directional guidance and the data obtained from the second star would provide the distance reference.

The Single Star Tracking System

In a single star tracking system an optical pickup or sextant is mounted so that it can be varied in elevation and azimuth in order to follow the star path. In some arrangements a tape recorded playback or program is used to supply information for continuously aiming near the star to be tracked. The timing for this tape playback is an important function because a star is at a particular angle with respect to a certain point on the Earth only at a particular instant of time. Actually, the tape programmer brings the star tracking system within the general direction of the star, and an error-detection circuit is required to center the telescope more exactly onto the star. The outputs of this system consist of several voltages that are proportional to the missile error in pitch, roll, and yaw.

After passing through the scanner, light from the star falls on the phototube causing a voltage output proportional to the light intensity. The phototube output passes through a selective amplifier which separates the signal from the noise. The desired signal is then detected to obtain the information of the error. In the first resolver the detected signal is resolved into azimuth and elevation error signals by comparison with a reference voltage from the scanner. The direction resolver can be a bridge demodulator circuit or some type of phase detector.

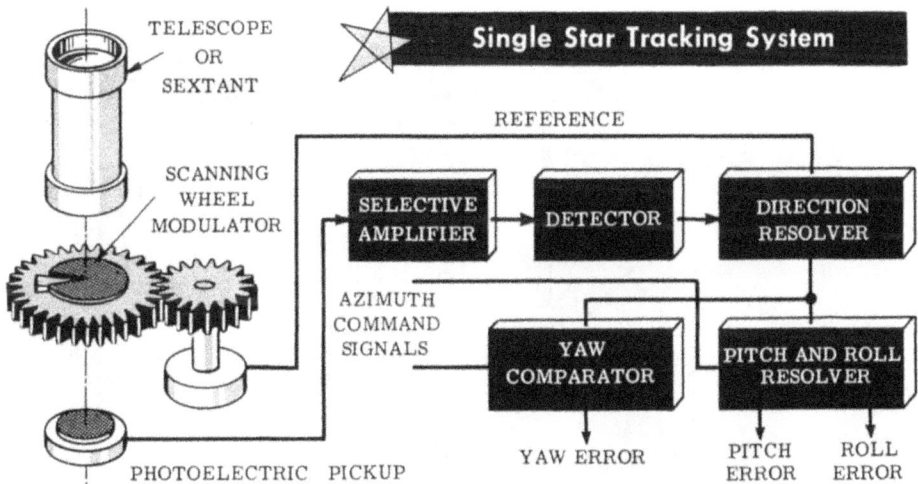

The errors are again resolved to make them usable in the missile's system of coordinates after the azimuth and elevation error in the sextant have been determined. The second resolvers are controlled from the tape reader. The same signal that sets the sextant sets a resolver for the elevation error output. If the elevation signal was not resolved in this manner, there would be no way of determining whether the error was in the pitch or the roll axis.

The Astro-Inertial System

If the telescope were elevated and pointed directly forward along the missile heading, any elevation error signal output would be attributed to error about the pitch axis. Also, if the telescope were pointed out of the side in the lateral direction, any elevation error would be only a function of roll. Therefore, the resolver is needed to determine whether the elevation error signal is due to pitch or roll, or a combination of the two.

A system better than a pure automatic celestial navigator can be realized by combining it with an inertial navigator. For instance, the inertial system can be set to provide guidance in a desired path to the target without reference to any associated information. Then, the stellar references simply aid the accelerometer and gyro combination to readjust itself for any drift or other deviations. This is certainly better than depending on finding the right star paths at the right time to give a complete guidance from those sources. Furthermore, an inertial system will operate under cloud cover or during other periods when the star references may not be available. However, in a cruise type trajectory the star information can be fed in whenever available to give data for corrections of the course which might be desirable.

An inertial system combined with a celestial tracker is known as a stellar supervised inertial autonavigator as compared with a basic continuous star tracker, which is referred to as an automatic celestial navigator.

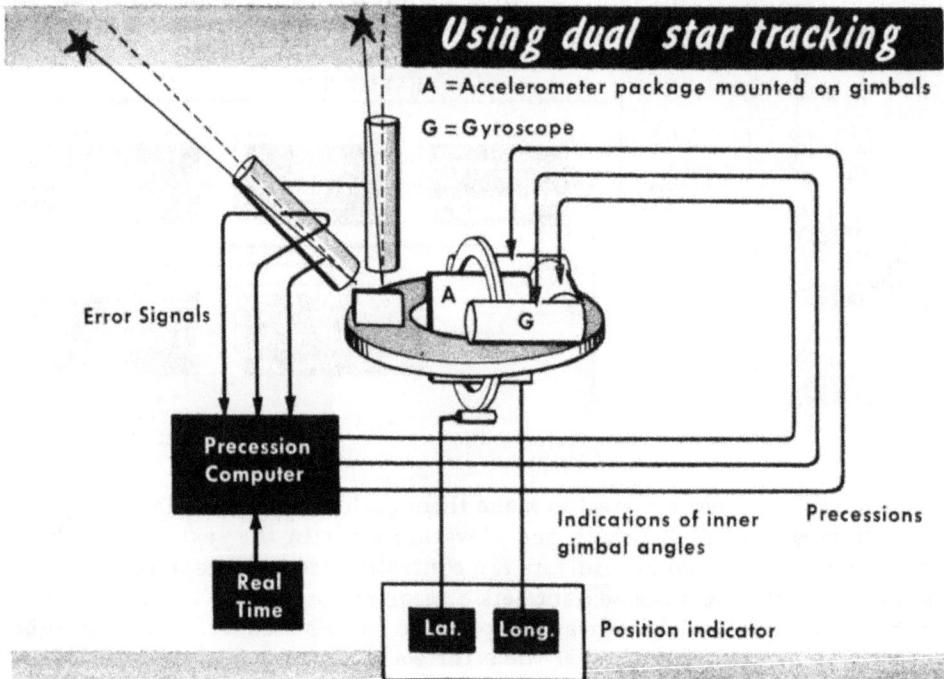

Using dual star tracking

A = Accelerometer package mounted on gimbals

G = Gyroscope

Error Signals

Precession Computer

Real Time

Lat. Long.

Indications of inner gimbal angles

Precessions

Position indicator

COMPUTER APPLICATIONS

Fire-Control and Bomb-Sight Applications

One of the earliest applications of computers to a military situation was developed in gun-fire control systems. World War II saw applications of both surface and antiaircraft fire control techniques. Early in the war naval forces utilized such systems very effectively against surface vessels. Either an optical or radar tracker, mounted well above the deck, located and followed the enemy ship. It established the range, bearing, and elevation of the target, and fed this data to the computer inside the ship's plotting room. By combining these quantities the line of sight to the target was established and the director followed any movement of the enemy ship. Our ship's direction and speed also had to be fed into the computer to establish a base for computation.

A stable element corrected for the roll and pitch of the ship. Ballistic data on the shell's time of flight, trajectory corrections, and drift also were fed to the predictor, where sight-angle and sight-deflection were established. By suitably combining all of these factors with wind speed and direction, train and elevation control was made available to the guns to enable them to fire in the proper direction to intercept the moving target.

Anti-aircraft fire control followed somewhat similar steps except that the speed of the target demanded more rapid computations and gun control.

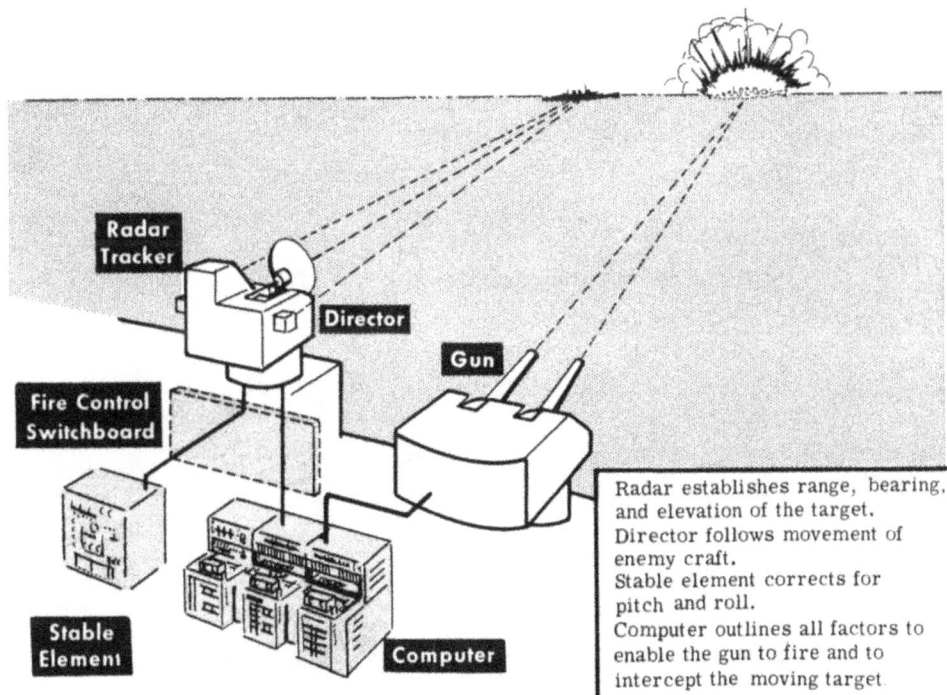

Radar Tracker
Director
Gun
Fire Control Switchboard
Stable Element
Computer

Radar establishes range, bearing, and elevation of the target.
Director follows movement of enemy craft.
Stable element corrects for pitch and roll.
Computer outlines all factors to enable the gun to fire and to intercept the moving target

COMPUTER APPLICATIONS

Fire-Control and Bomb-Sight Applications (contd.)

Aerial bomb-sight applications utilized the results of another early computer application — the determination of the correct range at which the bomb should be released. This distance is based on a number of factors. The time t required for the bomb to fall to the target is based on the formula, $t = \sqrt{2D/g}$, or $D = \frac{1}{2} gt^2$, where D is the distance to the target and g is the acceleration of gravity — approximately 32 ft per sec.

To determine the slant range at which the bomb should be released, it is necessary to calculate the vector sum of the bomber's altitude, A, and the distance between the release point and the target at ground level, R. The ground release range in a vacuum is $R = V_c t_t$, where V_c is the net closing speed of the aircraft and t_t is the time of fall of the bomb. Thus, the formula for the slant release range is $S = \sqrt{A^2 + (V_c t_t)^2}$ in a vacuum. Since the bomb will be falling through air, its range will be shortened by a distance known as the trail T, so in the air the slant release range, $S = \sqrt{A^2 + (V_c t_t - T)^2}$.

To solve the equation for slant release range the altitude, A, and the ground release range, $(V_c t_t - T)$, may be combined vectorially in a mixer-amplifier or by other means for obtaining the vector sum. If more than one bomb is to be dropped by the bomber, a factor known as *lead* is added to the ground release range. Wind velocity and direction must also be fed to the computer.

During the bomb run the operator observes the target either by an optical or radar sight and releases the bomb at the computed slant release range. During the period prior to the release the bomber is kept in stable flight by an autopilot and the sight is gyro-stabilized.

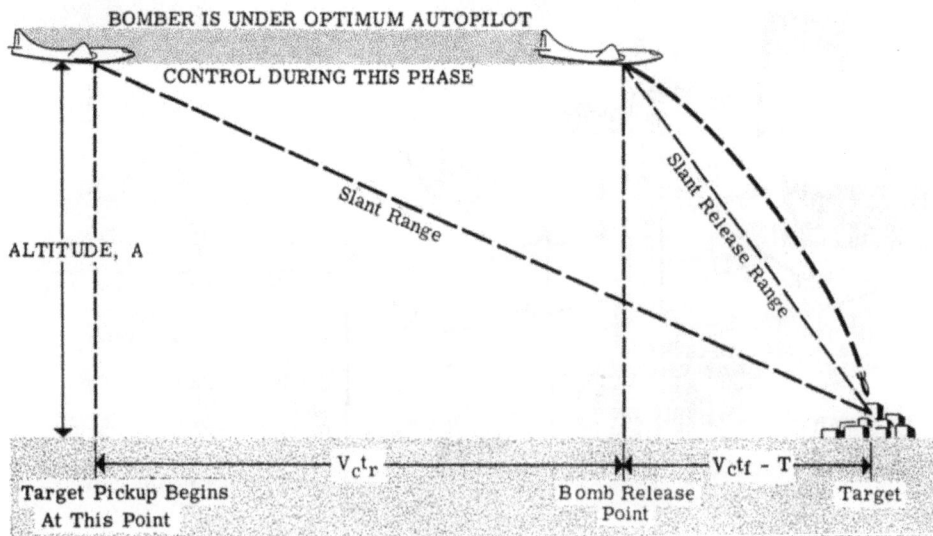

BOMBER IS UNDER OPTIMUM AUTOPILOT CONTROL DURING THIS PHASE

Slant Range

Slant Release Range

ALTITUDE, A

$V_c t_r$

$V_c t_f - T$

Target Pickup Begins At This Point

Bomb Release Point

Target

Command-Guidance Computers

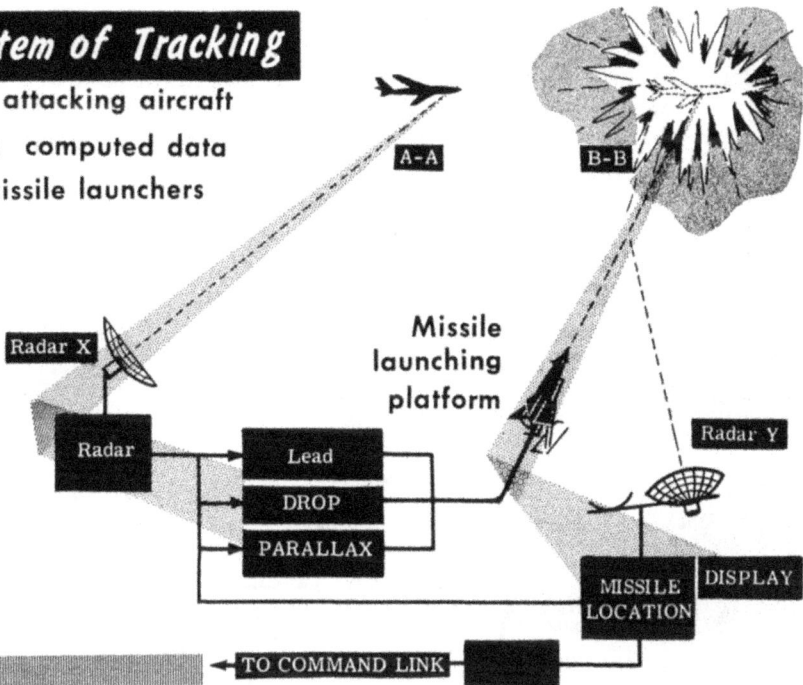

A System of Tracking
a single attacking aircraft
provides computed data
to the missile launchers

Surface-to-air missile guidance systems employ similar techniques to those used in antiaircraft gun-defense arrangements. In one of the first command-guidance systems for projector-launched defense missiles, a tracking radar (X) located the incoming bomber (A-A), just as in the case of the antiaircraft gun-defense technique. Another radar-signal receiving antenna (Y) tracked a beacon in the surface-to-air missile. To align the missile-launcher platform and the missile-tracking antenna properly, a computer capable of calculating the corrections necessary for lead angle, missile drop from a straight line trajectory, and parallax was employed. With this data the launching platform could be properly aimed and the operator at (Y) could follow the missile developing the proper command signals for a certain degree of accuracy. Additional computer output provided data to orient the missile-tracking antenna into the correct position relative to the roll-axis of the weapon. A missile receiver picked up the guidance commands.

The concept of tracking a single attacking aircraft and providing computed data to missile launchers is fundamental to any ground-to-air system. However, modern defense systems must embrace a much more elaborate arrangement to cope with large numbers of attacking aircraft, and provide direction to batteries of defensive weapons to be launched simultaneously against mass intrusion.

Computers in Flight Control

The advances in airborne computers for present-day navigation and weapon system control from the early bomb-sight computers have been equally spectacular. Through the years since World War II flight control and airborne military weapons systems have shown a trend toward an increasing degree of automaticity with the pilot involved more and more as a monitor of visual data displays and a selector of operational modes. In a typical airborne computer for high-speed aircraft, an air-data computer accepts airspeed, altitude, temperature, and angle-of-attack data to establish cruise, altitude,

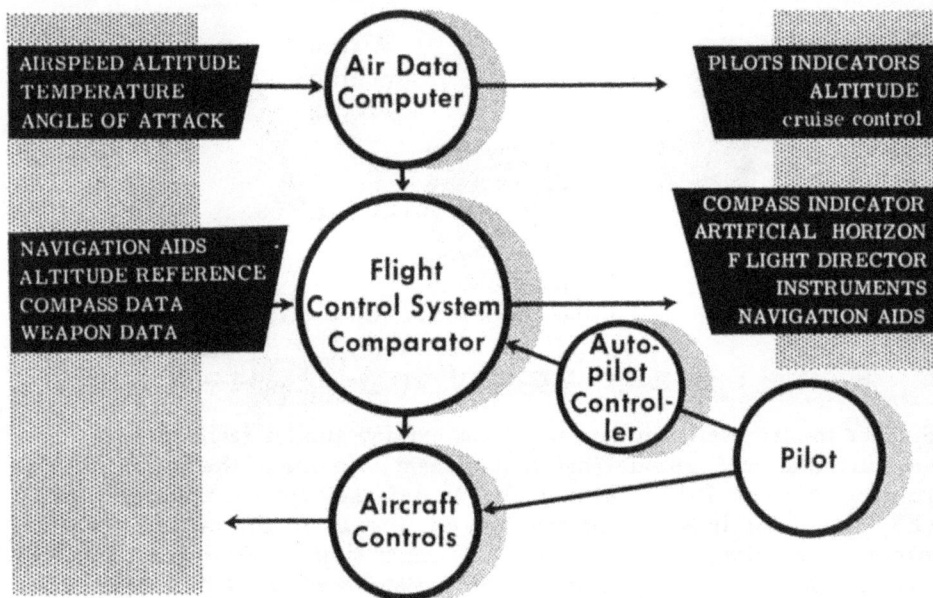

and tactical control. This data is fed to the main flight-control computer which combines it with navigational, attitude, compass, and weapon data. Its output supplies the necessary signals to operate the aircraft controls, and to establish the required flight path. Several outputs are provided for visual monitoring by the pilot so that he can take over if the occasion demands it.

In an evolutionary process toward pilotless aircraft and cruise-type missiles, which must navigate over considerable distances around the globe, the computer approach outlined here has been perfected to a further extent. As we have seen in earlier chapters, many missiles do not fly as aircraft but are more in the rocket category. Airborne, ground, and shipboard computer requirements vary considerably with the type of missile and guidance involved. However, in every case the computer must solve the problems necessary to missile stability and direction against either fixed or moving targets.

COMPUTER APPLICATIONS

Basic Computer Operations

In our consideration of the various sensing devices, we have seen the need for computers which combine signals in various proportions to feed through to the controller-amplifiers and actuators the resultant control signals to realize the proper ballistic trajectory. We considered sensors whose primary function involved *rate data*, and have seen the need for establishing rate signals suitable for feedback to prevent the over-controlling of thrust directors.

We indicated, in discussing inertial guidance, the need for *integrating* operations to convert acceleration data over a period of time into velocity and integrations of velocity to obtain a measure of the traveled distance. These computer functions apply to either control or guidance systems, and can be summarized into the following general categories: (1) Combining of separate signals by *addition, subtraction, division,* and *multiplication.* (2) Rate determination or *differentiation.* (3) *Integration.*

The simple addition of signals may be accomplished electronically with circuitry which normally is not thought of as computer networks. For instance, two or more signals can be combined at the *input* grid of a single tube by feeding them either in parallel or in series.

ADDING SIGNALS ELECTRONICALLY

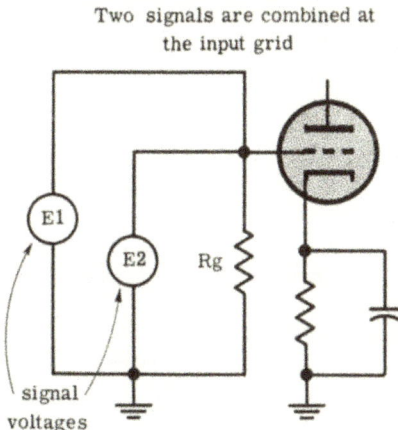

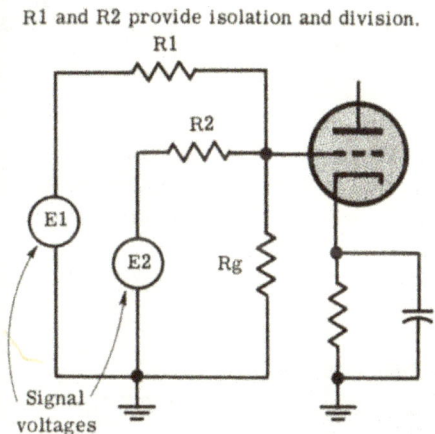

Two signals are combined at the input grid

R1 and R2 provide isolation and division.

They may also be fed in varying proportions and with some isolation between them by placing series resistors between the source and the grid resistor, R_g. Voltage division will take place between the series resistors and R_g to adjust the amount of signal accepted from each source.

COMPUTER APPLICATIONS

Basic Computer Operations (contd.)

Signals E1 and E2 may also be combined through a transformer, by connecting them at the opposite ends of a primary winding in the proper phase so

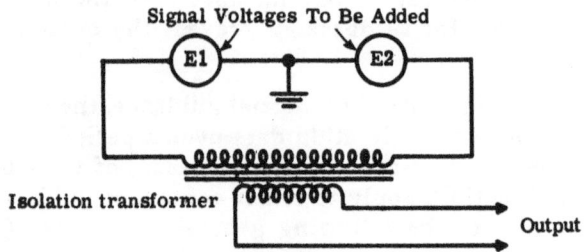

that the secondary voltages will add. If further isolation is required between the sources, they may be isolated through separate amplifiers.

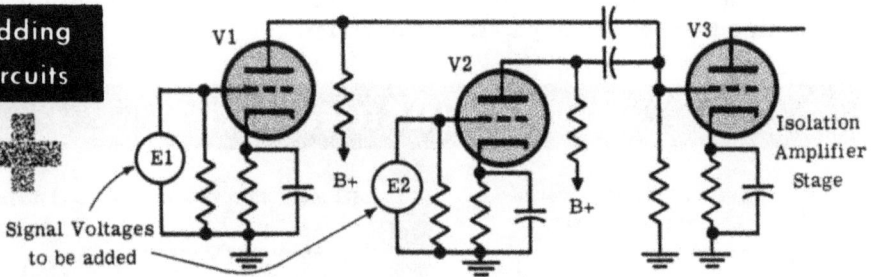

Voltages may also be subtracted by similar simple circuits, such as a center-tapped transformer primary winding connection, which will shift the phase of one input by 180° from the other. Such subtraction may also be accomplished with simple vacuum tube circuits by feeding two grids in opposite phase and combining their outputs in such a manner that a difference voltage is obtained.

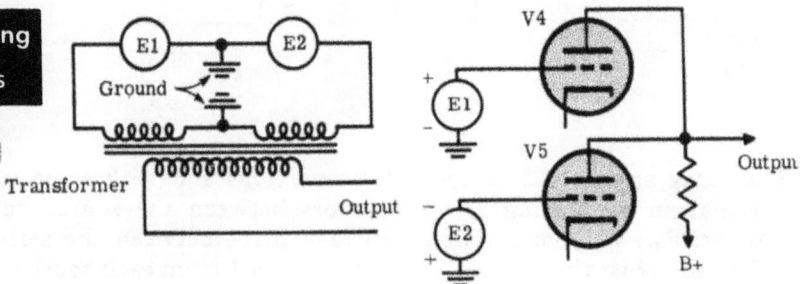

Basic Computer Operations (contd.)

Although we do not think of such circuits as normally belonging in the computer category, it is apparent that they are when compared to mechanical arrangements which *add* and *subtract*. For many years we have seen examples of mechanical devices, such as the slide rule, which add numbers mechanically to obtain a resultant summation. Also, quantities may be combined mechanically by lever or gear arrangements which either add or subtract revolutions. For instance, the inputs along two rods can be combined by a pivoted lever output as shown.

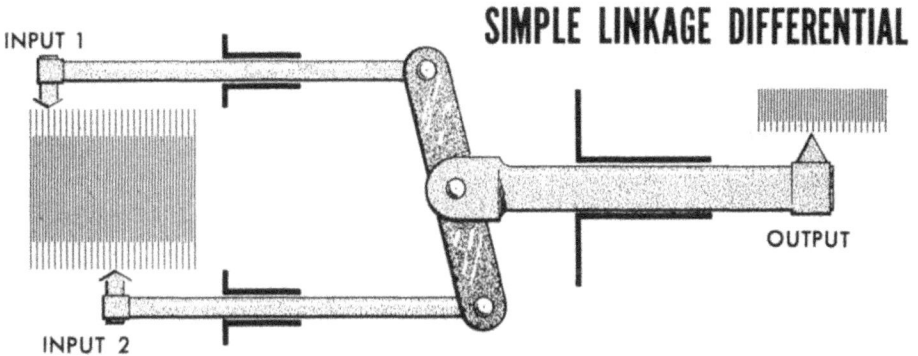

SIMPLE LINKAGE DIFFERENTIAL

INPUT 1

OUTPUT

INPUT 2

Typical of the geared system is the differential, which can add and subtract two quantities. It has two mechanical inputs and one mechanical output. The differential gear determines the resultant combination of two input quantities fed to gears on each side of the main bevel-gear arrangement.

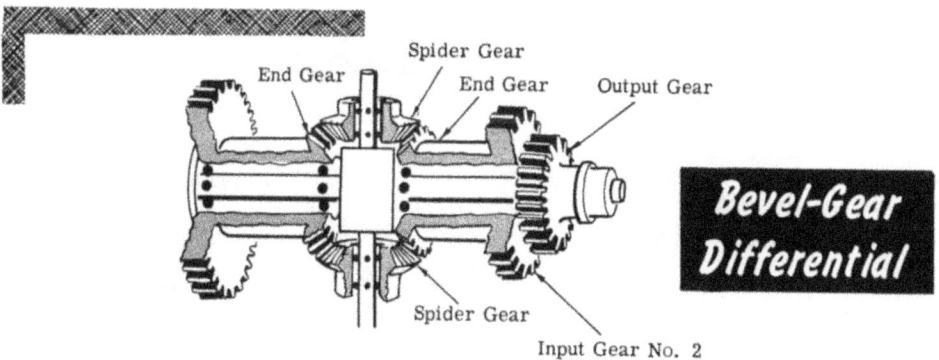

Spider Gear

End Gear   End Gear   Output Gear

Bevel-Gear
Differential

Spider Gear

Input Gear No. 2

Before we examine additional mechanical and electronic techniques, it is well to recognize that all computers are either of the analog or of the digital type.

Analog Computers

Analog types represent data by analogous quantities (for instance, an angle of rotation, a measured scale, a voltage). A speedometer indicates the velocity of an automobile by reading a quantity which is proportional to the speed of the car's drive shaft. Other examples are the slide rule and the differential. The slide rule multiplies, divides, raises to powers, and extracts roots by adding and subtracting logarithms respectively. The differential is used in automobiles, and also has been widely employed in gun

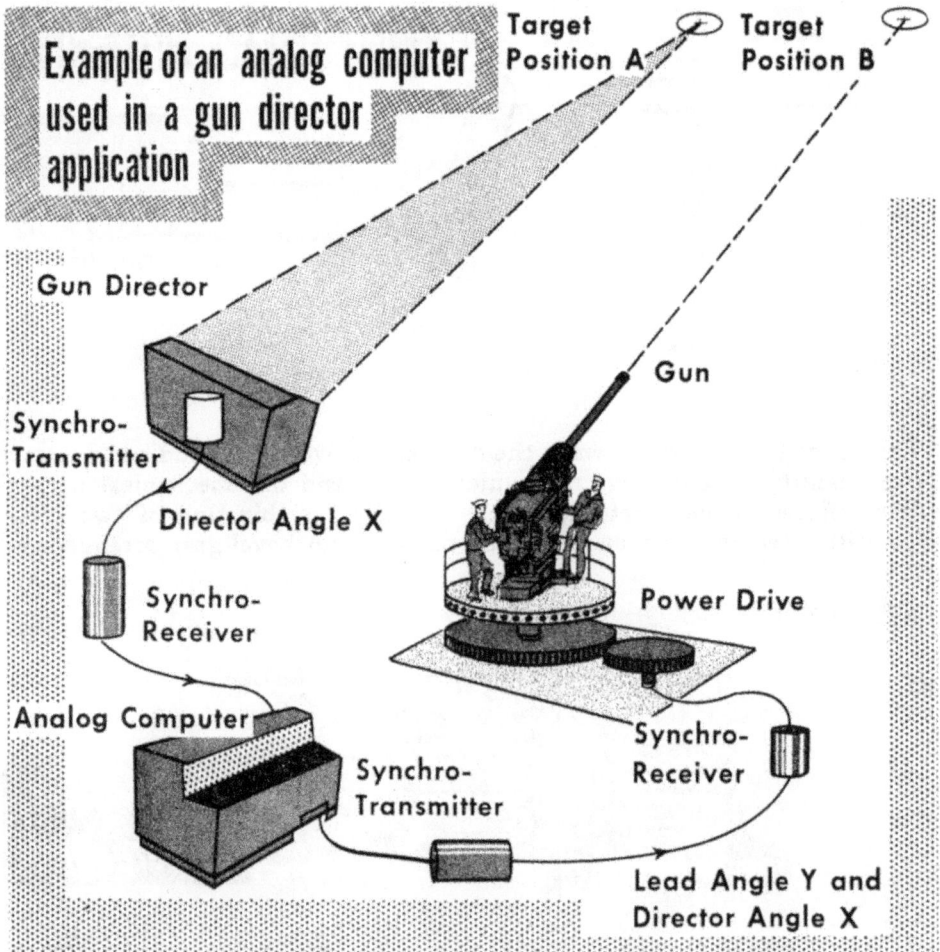

Example of an analog computer used in a gun director application

Target Position A

Target Position B

Gun Director

Synchro-Transmitter

Director Angle X

Synchro-Receiver

Analog Computer

Synchro-Transmitter

Gun

Power Drive

Synchro-Receiver

Lead Angle Y and Director Angle X

director applications. Mechanical analog-computer techniques were developed rather extensively during World War II to position guns correctly for firing at moving targets both at sea and in the air.

Analog Computers (contd.)

In addition to adding angles it is often necessary to integrate quantities in computers. A typical mechanical integrator is the *ball and disc* type, which has been used in bomb sights and in analog computers for missile applications. The position of the friction ball from the center of a rotating disc

Ball and Disc Integrator provides range information from the velocity output of an integrating accelerometer

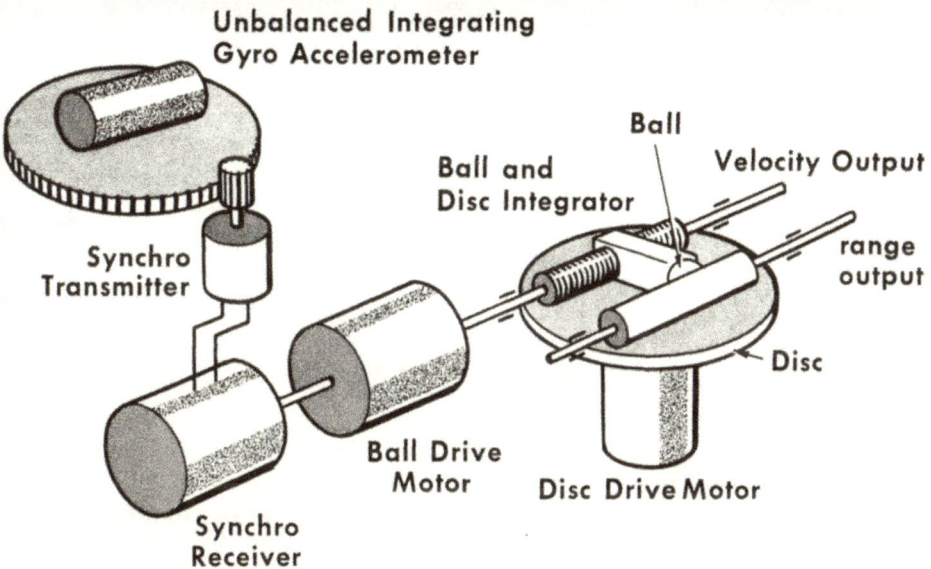

Unbalanced Integrating
Gyro Accelerometer

Ball and
Disc Integrator

Ball

Velocity Output

range
output

Synchro
Transmitter

Disc

Ball Drive
Motor

Disc Drive Motor

Synchro
Receiver

determines the rate at which an adjacent shaft rotates as the disc spins below the ball. The position of the ball is determined by the input applied along its rotating shaft. The number of revolutions of the shaft is a measure of the summation of the changes of the applied signals or their integral over a period of time. The output of a selsyn which is geared to the rotating shaft provides a signal for controlling the missile or for performing other functions in proportion to an integrated quantity — such as distance which equals velocity multiplied by time. Other examples of mechanical analog computers are certain types of gyroscopes that can give an integrated output if accelerations are applied to them.

Analog Computers (contd.)

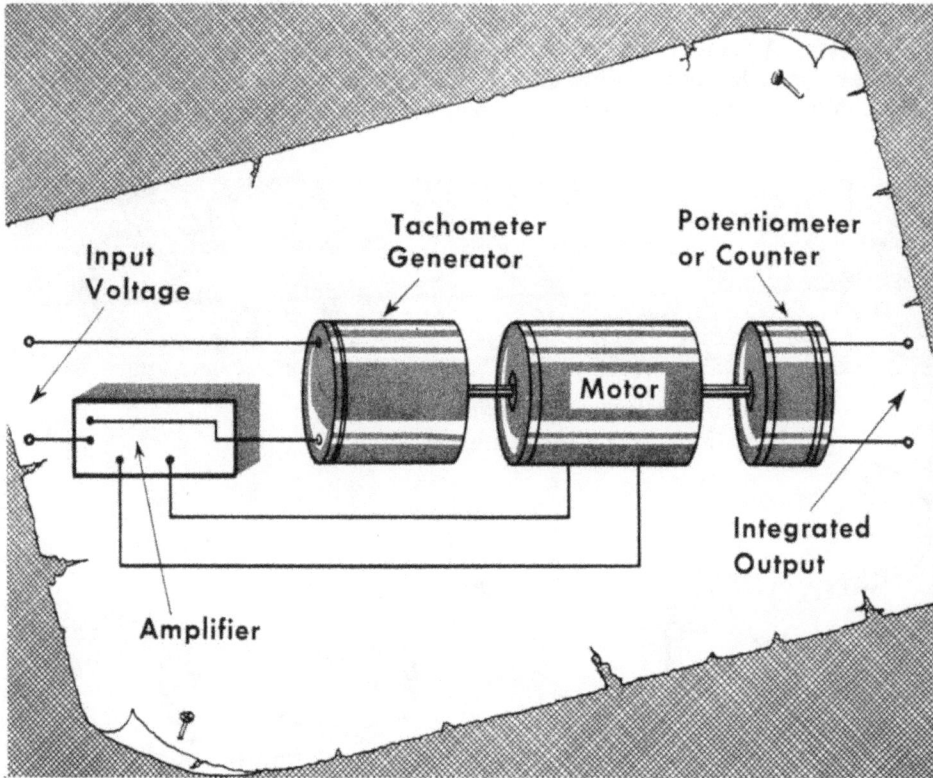

Although mechanical analog computers were developed first, it soon became apparent that combinations of electronic and electrical components along with mechanical ones could do the job better than pure mechanical components. A typical arrangement of this type utilizes a motor-driven tachometer-generator, an amplifier, and some means for recording the rotation of the driving motor. If we desire to integrate the level of a given input over a period of time, it can be fed into the amplifier along with the voltage output of the motor-driven tachometer-generator so that a balanced or zero-input is obtained. Any change in the input signal will result in a change in the amplifier output which will cause the motor to rotate the tachometer-generator to again establish balance.

A counter on the end of the motor shaft will record a summation of the motor rotations over a period of time or their integration which is directly proportional to the integration of the input signal against which equal balancing signal voltages were generated. This counter may be either mechanical or a rotating potentiometer producing voltage changes, which can be added.

Analog Computers (contd.)

The use of electronics in analog computers has been extended to provide much higher operating speeds than those of pure mechanical types, with outputs which are more compatible with the electrical, electrohydraulic, or electropneumatic actuators controlling the missile-thrust directors. Since analog computers do not count as do the digital types, and since they utilize analogous quantities, addition and subtraction of voltages are often carried out in the simpler adder and subtractor circuits previously mentioned. Differentiation and integration are performed by resistance, capacitance and/or inductance combinations. Integration may be realized by charging a capacitor through a series resistor, and using the voltage drop across the capacitor as the output. If the voltage across the resistor is utilized, differentiation results because only the rate of change during the charging period is utilized.

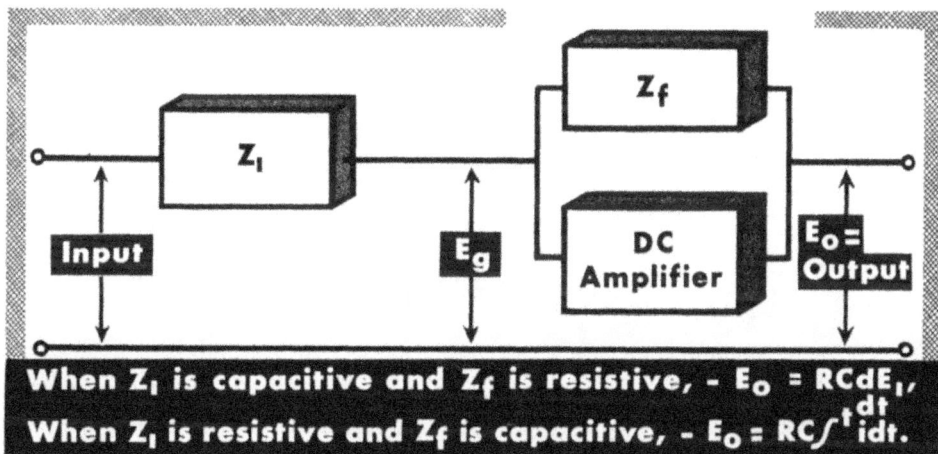

When Z_1 is capacitive and Z_f is resistive, $-E_o = RC\dfrac{dE_1}{dt}$,

When Z_1 is resistive and Z_f is capacitive, $-E_o = RC\int^t i\,dt$.

A circuit common to all electronic analog computers is the d-c feedback amplifier. By using resistance or capacitance for $Z1$ and Zf, it is possible to reverse the sign of the input voltage, to multiply by the ratio of Zf to $Z1$, to differentiate, and to integrate. If $Z1$ is capacitive and Zf is resistive, differentiation occurs at the output. If $Z1$ is resistive and Zf is capacitive, integration is obtained from this arrangement. It is beyond the scope of this book to describe the detailed circuitry.

D-c amplifiers suffer from drift, and require compensating circuits of special design as well as well-regulated power supplies to minimize these tendencies. Vacuum tubes are often aged to reduce drift. Balancing of the output voltage to zero for zero input voltage is standard procedure in many analog circuits. D-c feedback amplifiers generally have high input impedance, very high gain tubes, and a low output impedance.

COMPUTER APPLICATIONS

Analog Computers (contd.)

Since analog computers will perform all of the basic mathematical operations mentioned earlier in this chapter, they are employed in many missile applications. They have played a large role in the simulation of missile flight trajectories and guidance problems, permitting designers to *fly* the missile in the laboratory before it is put into expensive hardware.

Analog computers have been widely used in high-speed aircraft to perform all of the various necessary flight and weapon-data calculations. They have also been used in many missile applications, where differentiation and integration operations are required. They have offered a weight advantage over digital computers due to some extent to the fact that much of the input data has been in analog form from the sensors, and that converters are not required. However, for many applications the trend is away from them, because they are limited in accuracies to about one-tenth that of digital computers. This situation is rather obvious when we again recognize that they do not count but simply combine analogous quantities.

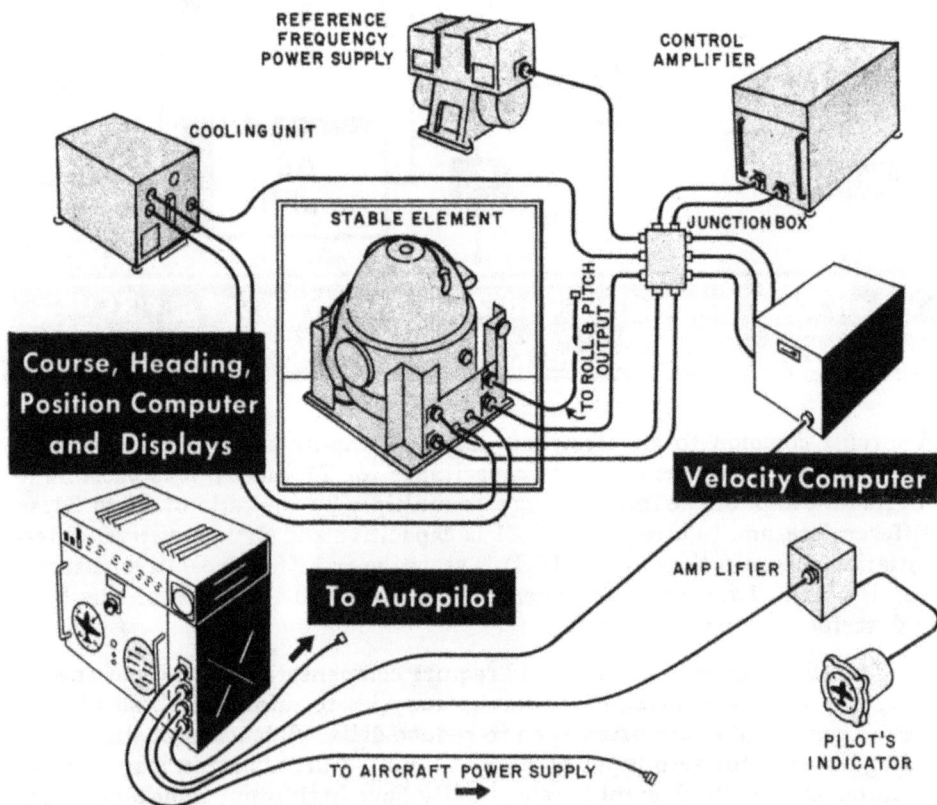

REFERENCE FREQUENCY POWER SUPPLY

CONTROL AMPLIFIER

COOLING UNIT

STABLE ELEMENT

JUNCTION BOX

TO ROLL & PITCH OUTPUT

Course, Heading, Position Computer and Displays

Velocity Computer

To Autopilot

AMPLIFIER

TO AIRCRAFT POWER SUPPLY

PILOT'S INDICATOR

COMPUTER APPLICATIONS

Digital Computers

In missile applications it is often necessary to relate the difference between two large numbers, such as the difference between the target and missile positions when they are some distance from the launching site, or to compare the actual trajectory with the programmed one for ballistic types. When the distance from the launching site becomes great, the need for accurate calculations becomes more acute. To obtain the highest precision it is desirable to utilize digital techniques, but often the accuracy of the other components are such that the analog technique can be just as satisfactory.

Basically the digital method is as old as man's ability to count his fingers and toes. Another long-known elementary digital device is the abacus, which extends man's counting capability. However, modern electronic digital computers make possible rapid solutions of complex problems which involve large numbers of digits carried out to several places. By utilizing a large number of tubes, transistors and/or magnetic cores in counting, gating, switching, and memory operations, an unusually large amount of data can be handled at a high rate of speed.

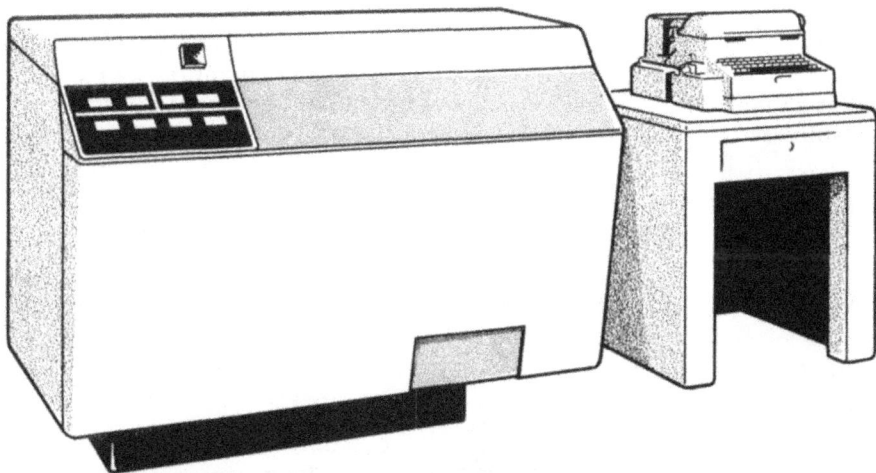

Missile data reduction is handled by digital computers which accept, analyze, and store a wide variety of test and operational data.

COMPUTER APPLICATIONS

The Analog-to-Digital Converter

Since the data from mechanical sensors is often in analog form rather than digital form, it is necessary to convert analog data to digital data. Although electronic analog data can be converted by circuitry and switching devices, a typical electromechanical analog-to-digital converter involves a rotating disc and a photo-electric pickup. The light beam passing through the disc at the transparent points will produce an output of separate pulses in digital form.

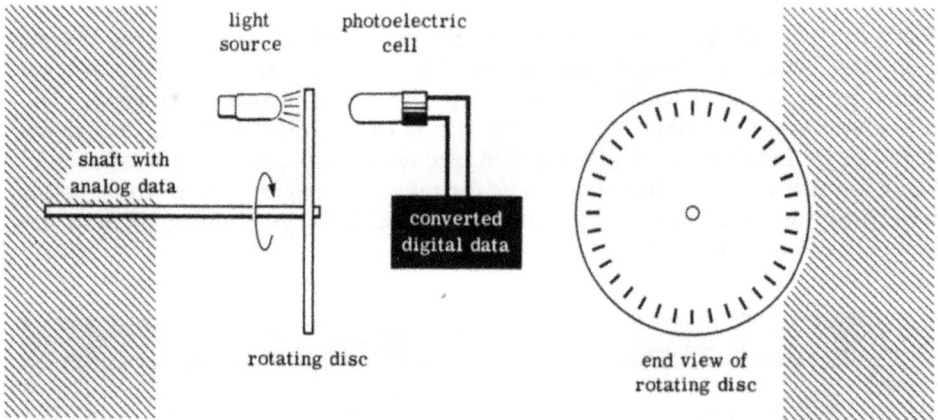

Another analog-to-digital conversion method utilizes rotary phase-shifting transducers to shift the phase of pulse generator signals in proportion to the rotation of the transducer. The output of a digital computer can be converted to analog form by a reversed procedure involving phase detectors and phase shifters to position a motor by an amount proportional to the binary count.

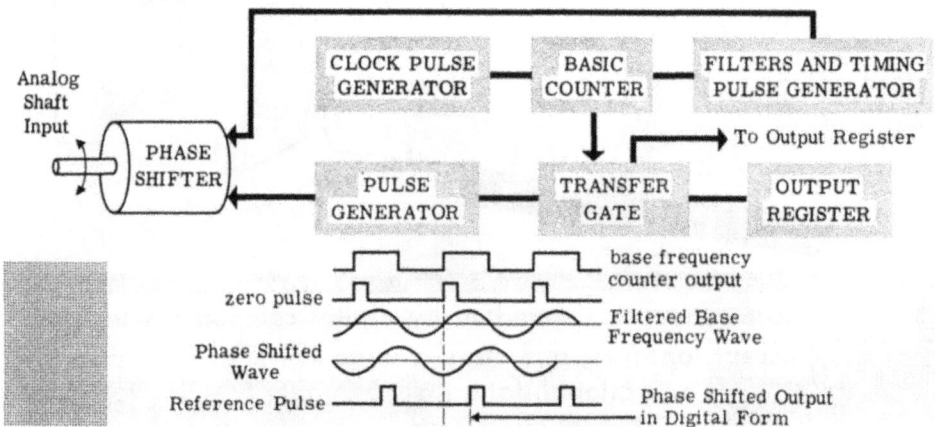

Binary Arithmetic Operations

Because of the *on-off* character of electronic switching, digital computer counting is generally done in binary form. Thus, the computation is done in terms of *1* or *0*. In other words, there is a quantity present or none is present. Such a technique is known as a binary count. The relationship of decimal and binary numbers is shown here. In a binary system, a sequence of zeros and units indicates the presence of the number 2 raised to some power. For instance, 111 indicates $(1 \times 2^2) + (1 \times 2) + 1 = 7$ or $1001 = (1 \times 2^3) + (0 \times 2^2) + (0 \times 2^1) + 1 = 9$.

Binary arithmetic operations are performed by combinations of several types of circuits — such as, so-called *flip-flops, and/or* and *counters*. These circuits may employ tubes, transistors and/or magnetic cores. A basic *flip-flop* circuit consists of two cross-coupled amplifiers arranged so that one drives the other.

The counting ability is derived from the fact that a negative pulse at the input causes a change from one state to the other. Cores can also be used to obtain *flip-flop* action.

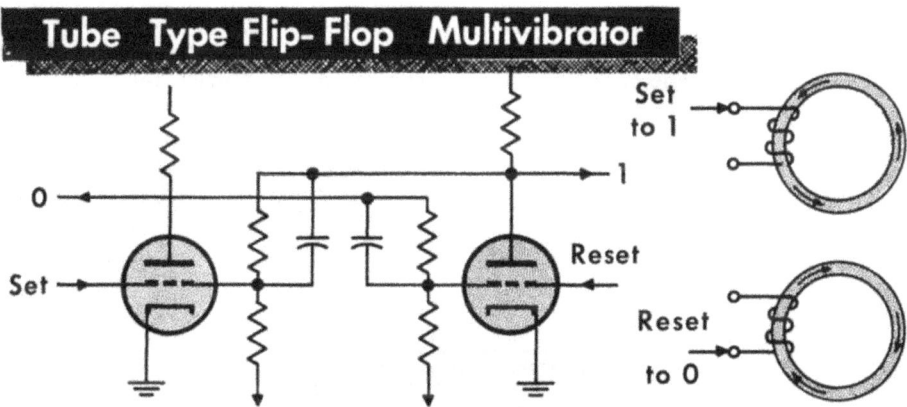

Tube Type Flip-Flop Multivibrator

Magnetic Cores also have similar characteristics

EQUIVALENT NUMBERS IN THE DECIMAL AND BINARY SYSTEM			
DECIMAL NOTATION	BINARY NOTATION	DECIMAL NOTATION	BINARY NOTATION
0	00000	6	00110
1	00001	7	00111
2	00010	8	01000
3	00011	9	01001
4	00100	10	01010
5	00101	11	01011

COMPUTER APPLICATIONS

Typical Digital Computer

A typical digital computer[2] for missile guidance consists of input, memory, control, arithmetic, and output sections. The input in this application involves error signals from sensor units arranged to provide the information necessary to determine the degree to which the missile is maintaining the desired course. The control circuits determine the type of computing operation to be performed. The program information from the memory section provides the control unit with instructions for its operation. The constant storage unit provides necessary reference information, which has been placed into the computer system before the guidance operation is undertaken.

The dynamic storage continually accepts new information and maintains a running memory of the operation. The arithmetic unit does the mathematical processing — such as, addition, subtraction, multiplication, division, and root extraction as well as related electromathematical operations. Vacuum tube or transistor counters, gates, relays, multivibrators, and other switching elements are utilized for the arithmetic operations. Magnetic cores are also included in some designs, but are used more widely in memory circuits. The output unit of a digital computer may consist of switches, relays, tape recording elements, automatic printers, and geared indicators. In the missile application it will most likely feed out information to drive amplifiers ahead of the controllers and actuators of the thrust deflectors.

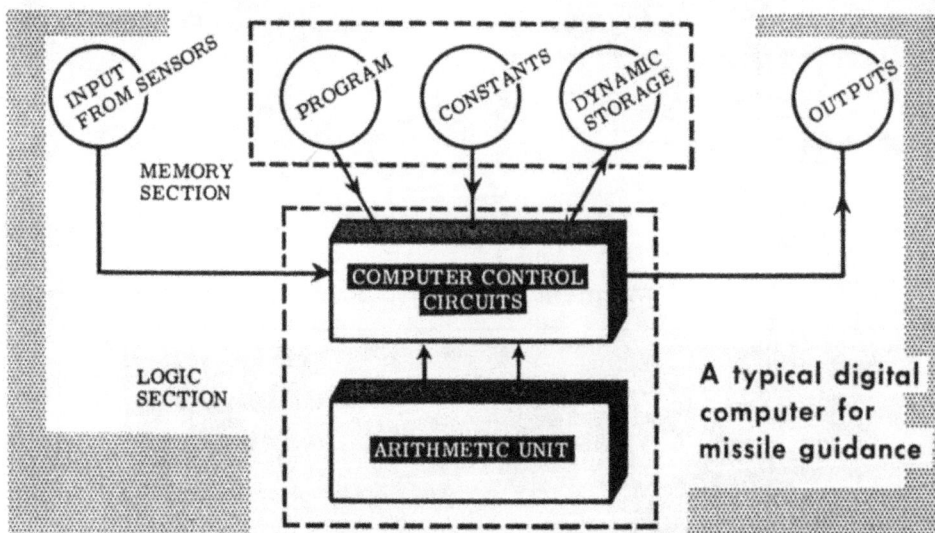

A typical digital computer for missile guidance

---

[2] Detailed descriptions of digital computer circuitry and many facets of their operation is beyond the scope of this book, but is dealt with thoroughly in *Basics of Digital Computers* by John S. Murphy: New York: John F. Rider Publisher, Inc., 1958.

Typical Digital Computer (contd.)

Although the control and arithmetic units (sometimes, referred to as the logical section) perform the basic calculations, the memory section has been called the backbone of the system. It must store certain computations, and provide them back to the logical unit as required. Considerable research and development effort has been put forth to make it possible to tap these memory elements over and over again, if need be to accomplish the complete computations.

Undoubtedly, the human brain is one of the most remarkable storage devices. With its 10 billion nerve cells it presents a capacity considerably beyond any electronic computer storage physically possible with present-day components. Magnetic drum storage has been widely used to place as much as a million bits of information on a compact cylinder. The technique is comparable to magnetic tape recording with provision for read-out as required. A photographic method of storage, which offers a capacity of about a billion bits of information and begins to approach the human brain, has been mentioned in the literature.

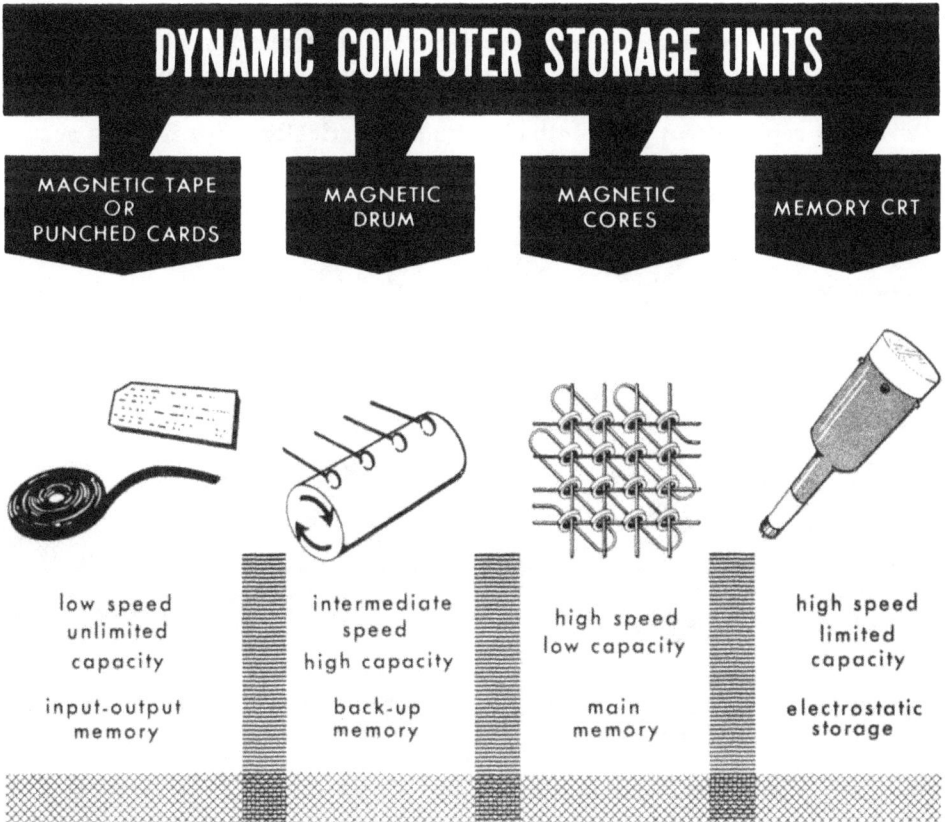

DYNAMIC COMPUTER STORAGE UNITS

MAGNETIC TAPE OR PUNCHED CARDS	MAGNETIC DRUM	MAGNETIC CORES	MEMORY CRT
low speed unlimited capacity	intermediate speed high capacity	high speed low capacity	high speed limited capacity
input-output memory	back-up memory	main memory	electrostatic storage

Typical Digital Computer (contd.)

A storage arrangement which provides high-speed operation and which is consequently of interest in missile application is a magnetic core arrangement. Coincidence of current in two of the coordinate wires gives sufficient magnetic force to produce remanent magnetic induction in the core at their intersection. Output is obtained from a third wire passing through each core. This system is free of moving parts and, although its power requirements are relatively high, it offers considerable compactness. Other types of well-known storage devices involve delay lines and electrostatic charges on the face of a cathode-ray tube. Programming and constant data may be stored in plugboards or wired matrices, but tape is often employed in more complex applications.

Although considerable progress has already been made in the application of transistors to computers in this country and elsewhere, the future offers promise of wider transistor application as their operating stability limits are raised. The use of magnetic cores together with transistors offers promise for a new order of compactness. It is obvious with the large number of individual components in a digital computer that reliability is a matter of great concern. However, even in computers employing only vacuum tubes, great progress has been made in the direction of higher reliability. While much of the progress has been made through designs which allow for safe margins of operation, part of the improved performance is realized by switching replacement parts into circuits automatically when failures occur.

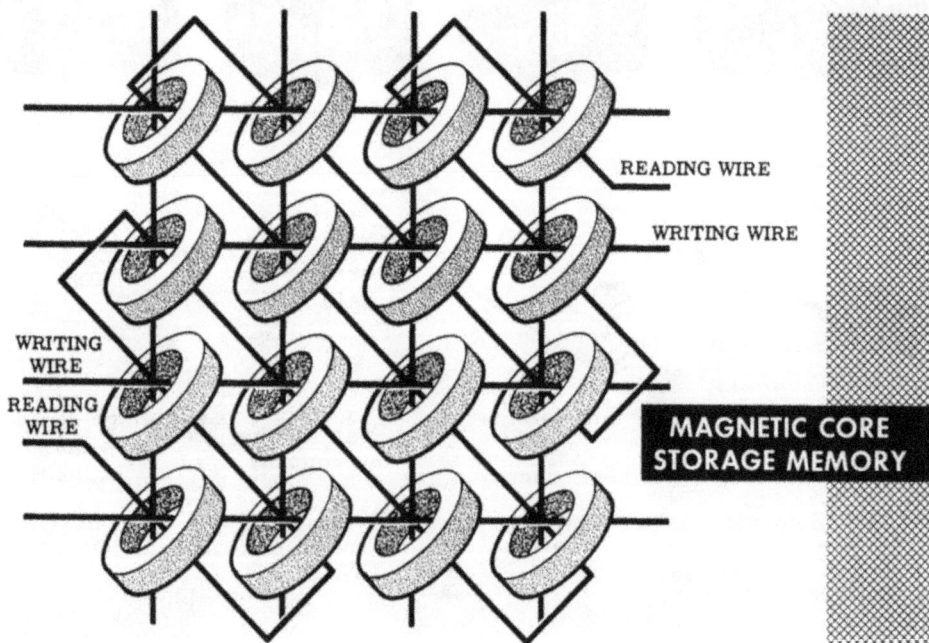

READING WIRE

WRITING WIRE

WRITING WIRE

READING WIRE

MAGNETIC CORE STORAGE MEMORY

ACTUATORS AND RELATED COMPONENTS

Introduction

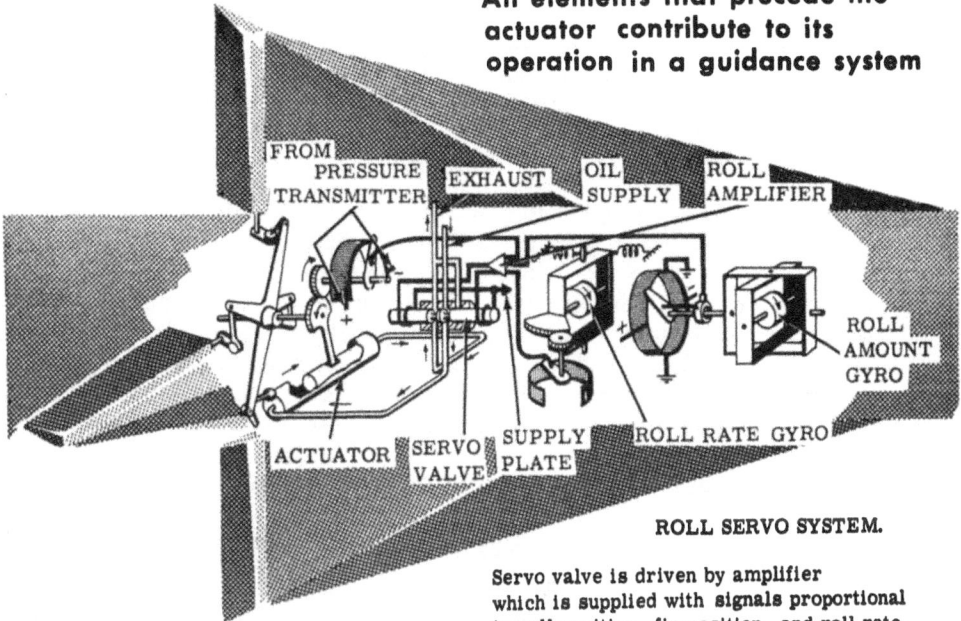

All elements that precede the actuator contribute to its operation in a guidance system

ROLL SERVO SYSTEM.

Servo valve is driven by amplifier which is supplied with signals proportional to roll position, fin position, and roll rate.

All of the guidance and control information in a missile is collected and developed for just one purpose: To supply the necessary input to the actuators of its surface- and thrust-control elements. The final objective of the control or guidance system is to move or actuate the control surfaces or thrust directors in such a way as to achieve the desired attitude or position of the craft. Thus, in a sense the actuator compares with the loudspeaker of a radio set because it is the output unit of the system. Everything that precedes it in the guidance system is manifest through it.

The escapement-controlled *rubber-band* motor-actuator of a model aircraft is a far cry from the hydraulic or pneumatic system of a guided missile. However, it demonstrates the basic principles very well, and provides enough power to control the lightweight surfaces of a very small craft. In the simple model the tension in the twisted rubber band causes the escapement arm to rotate the control rods leading to the rudder or elevator when a signal pulse is fed to the escapement coil. In the guided missile the guidance signal, whether from internal or external sensors, is fed to an amplifier which controls a valve in a hydraulic or pneumatic system. In the model craft electric motor control of the surfaces is also often employed to adjust their position. In this arrangement servomotors are caused to rotate in either direction by input from amplifiers ahead of these actuators and between them and the guidance information source.

ACTUATORS AND RELATED COMPONENTS

Electric Motor Control

For missile guidance a most obvious approach is to utilize electric motors to drive the thrust-directors or the control-surfaces. However, a very large and powerful motor would be required to move these elements if direct drive was employed. By using a small motor, running at high speed, and geared down to the proper speed a practical arrangement can be realized. Either constant-speed motors operated through clutches or variable-speed motors are used.

Of course, there are advantages and disadvantages of electric motor-control. A major disadvantage is the fast rotation of the motor which is required to correct any sudden course deviations of the missile. It is difficult to overcome the large amount of inertia which exists in trying to suddenly reach a high-rotor and gear-train speed from a standstill. The lag introduced by this problem can easily be sufficient to make the system ineffective. Even if the weight of a large motor could be tolerated, its mass would still leave one with the inertia problem. However, two partial solutions to this problem have been realized by the development of small high-torque motors and by the use of a continuously operating motor through a clutch drive.

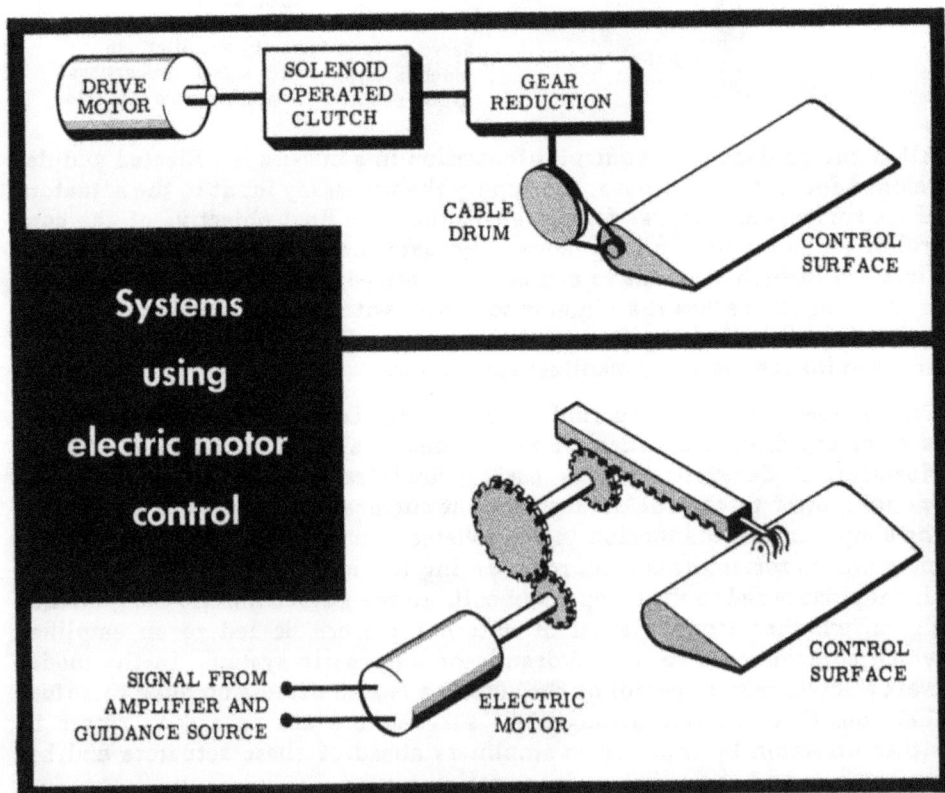

Electric Motor Control (contd.)

When a variable speed motor is used directly without a clutch, it rotates the driven element in either direction depending upon the sense of the controlling signal voltage. The motor rotates at a speed which is roughly proportional to the strength of the signal. It is obvious that a high-signal power input is required to supply the driving power of the motor.

When a constant-speed motor actuator is employed the motor runs continuously, and maintains a uniform rate of speed. A clutch provides the means for varying power transmission from the motor to the controlled element. The use of two clutches and a gear differential allows control in both directions. Due to their variable speed characteristics d-c motors are more suitable than a-c motors as missile actuators. Also, they generally develop higher stall torques, and are more suitable for handling heavy loads of the type encountered in high-speed missile control. There are two ways of controlling the speed and direction of a d-c motor in this application. One method is by controlling the voltage applied to the field, and the other method is by controlling the armature voltage. By controlling only the field voltage the armature rotation speed can only be increased. However, by also controlling the armature, the speed can be reduced to zero, and increased in one direction or the other depending upon the magnitude and phase of the voltage.

A typical d-c generator-motor arrangement is one in which the controlling signal is fed into the field winding of a constant-speed generator. The constant-speed generator provides an output amplitude of the proper phase to move the d-c motor in the required direction.

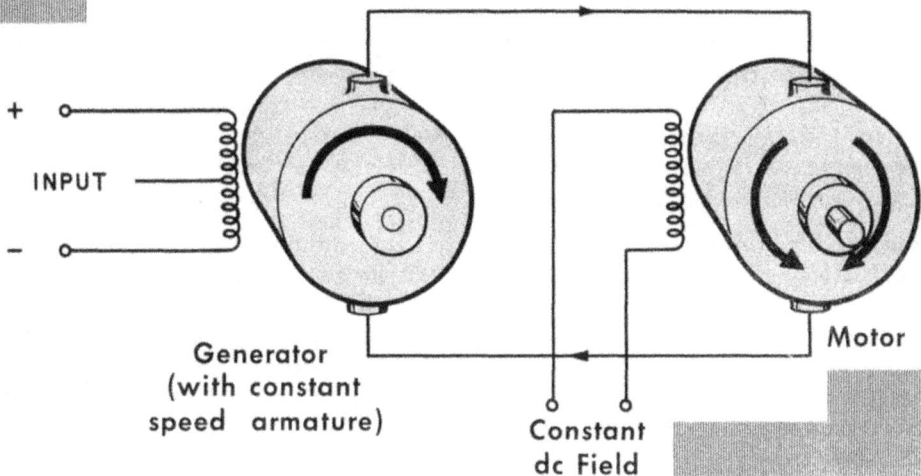

Generator (with constant speed armature)

Constant dc Field

Motor

INPUT

+

−

Pneumatics and Hydraulics

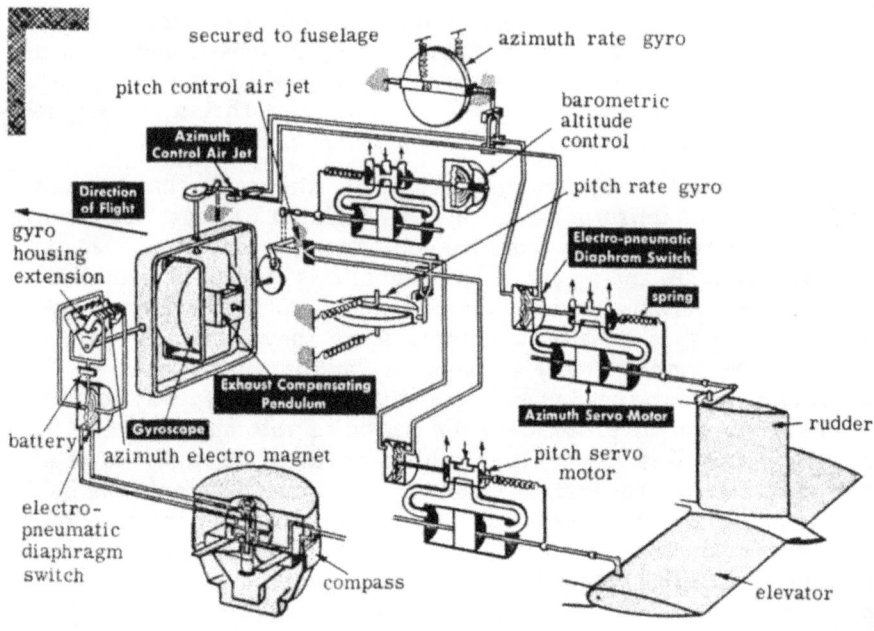

As mentioned above guided missiles require high speeds of response and large forces to move the control elements. Such requirements may be realized by either pneumatic or hydraulic systems. Although completely pneumatic systems have been employed in missiles, such as V-1 and the JB-2 pilotless aircraft, pneumatic and electrical means must be combined to be considered for modern weapons. A complete hydraulic solution is not practical in a guidance system either because it would require a hydraulic pickup from the gyroscopic and acceleration sensing-elements. Such an arrangement would exert so much force against the gyro elements that excessive errors would occur. It would also be necessary to develop a hydraulic computer. Such complete hydraulic systems appear to be practically impossible with present-day techniques and requirements. Therefore, as in the pneumatic systems a combination of electrical means and hydraulic means is essential.

Before going into detail regarding either hydraulic-electric or pneumatic-electric systems, first let us review some of the general hydraulic and pneumatic principles. Hydraulic systems are familiar to everyone in many commonplace applications, such as hydraulic brakes and transmission systems in automobiles, hydraulic lifts, and car jacks in garages. All such applications are based on Pascal's law which states that when a pressure is applied to a confined liquid, it is transmitted undiminished in all directions through the liquid; the shape of the container has no effect within reasonable limits.

ACTUATORS AND RELATED COMPONENTS

Hydraulic Systems

One of the major advantages of a hydraulic system or any other pressure system is that it eliminates the use of gears and other complex mechanical arrangements for the transfer of energy. As a result its reaction time is exceedingly rapid, since there are no series of mechanical backlashes or inertias to be overcome. By varying the surface area against which pressure is applied and for which the output is derived, it is possible to obtain a high degree of amplification in terms of the working force delivered.

For instance, it is possible to apply 100 pounds of pressure through an input piston and to transmit 2000 pounds or any greater or lesser amount of pressure against an output piston or lift. The amplification varies in direct proportion to the ratio of the output and input areas. The only factor which prevents this result from being the free ride that it might appear to be is that the larger area moves a smaller distance. In the above example, if the input piston moved 10 inches, the output piston would move only one-half inch. Actually, the volume of liquid remains constant, so one can always determine the output force and distance moved by recognizing that the output volume displaced equals to the input volume.

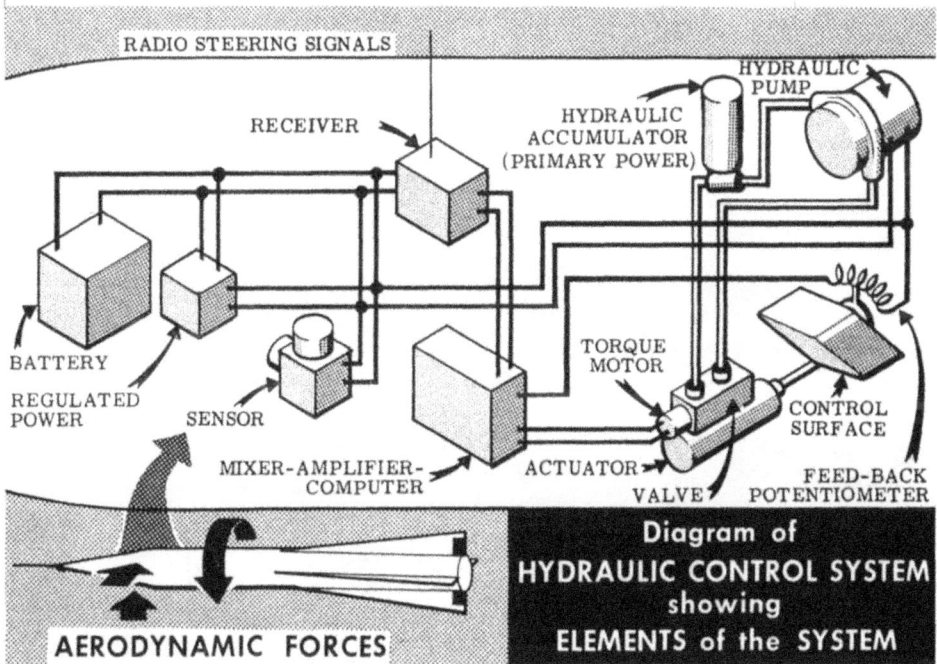

Diagram of
HYDRAULIC CONTROL SYSTEM
showing
ELEMENTS of the SYSTEM

Hydraulic systems are most widely used in missiles because they give high response control and maximum reliability. Also, they are light in weight and are compact considering their performance.

ACTUATORS AND RELATED COMPONENTS

Hydraulic System Elements

A typical hydraulic system contains a pump for bringing the liquid from a reservoir and pumping it through lines where its entry against an output piston is controlled by a directional transfer valve. Such a valve permits application of the liquid either above or below the piston and thereby permits operation in either direction. Pumps may be either of a geared type or piston type. In the former the liquid is forced between two tightly meshed gears as it travels from the input to the output point. A piston pump is simply what its name implies. Double-acting features together with a greater number of pistons result in greater and smoother fluid flow. To explain other parts of the hydraulic system, the reservoir is a storage place for the fluid used in the system. Pressure-relief valves are incorporated to limit the fluid pressure to some maximum value to prevent damage to the system. Generally they involve a spring-loaded block or ball which moves aside and allows liquid to escape back to the return line when the pressure in the feed line exceeds a certain amount.

Practically all modern hydraulic systems utilize oils as fluids. However, these liquids must be chosen carefully for their properties to provide satisfactory operation. Some of the factors of importance are chemical stability under continued use, freedom from acidity to avoid corrosion, lubricating power and viscosity or resistance to flow.

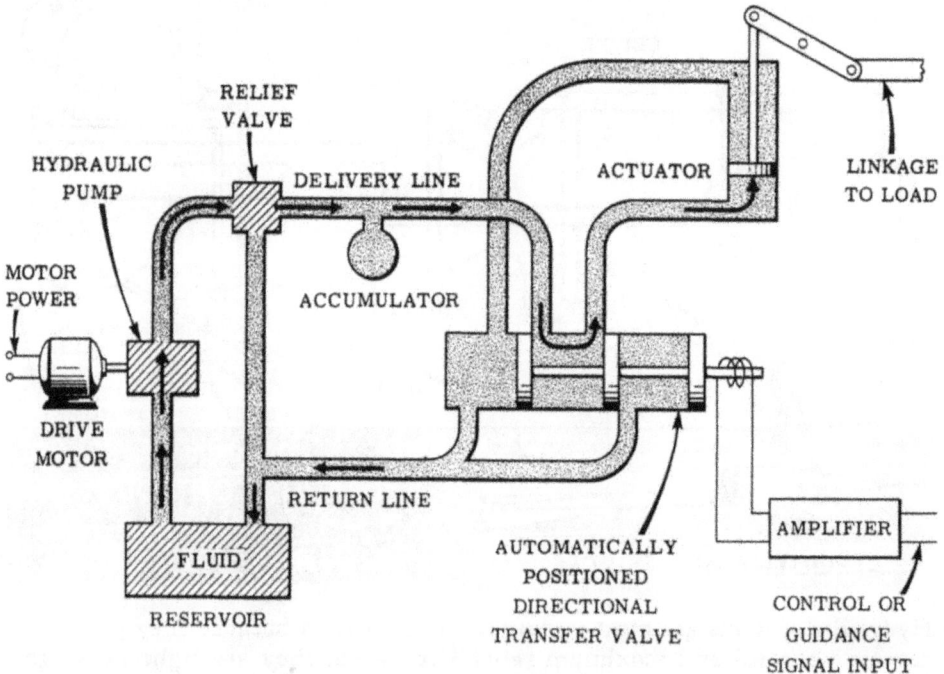

Valves, Relays and Motor Controllers

It will be noticed from the illustrations that either hydraulic or pneumatic electric systems incorporate solenoid-operated valves to regulate their operation. Such units are known as *controllers*. Although in some diagrams they are shown with single solenoid windings, they usually employ double windings so that a balanced action is obtained as the phase of the control

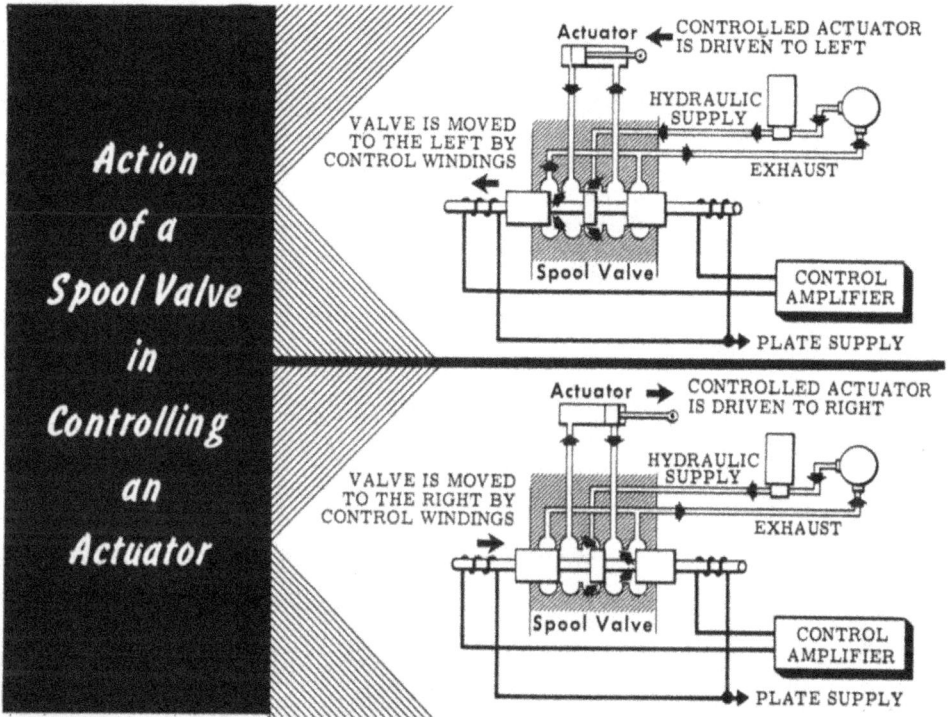

Action of a Spool Valve in Controlling an Actuator

Actuator — CONTROLLED ACTUATOR IS DRIVEN TO LEFT
HYDRAULIC SUPPLY
VALVE IS MOVED TO THE LEFT BY CONTROL WINDINGS
EXHAUST
Spool Valve
CONTROL AMPLIFIER
PLATE SUPPLY

Actuator → CONTROLLED ACTUATOR IS DRIVEN TO RIGHT
HYDRAULIC SUPPLY
VALVE IS MOVED TO THE RIGHT BY CONTROL WINDINGS
EXHAUST
Spool Valve
CONTROL AMPLIFIER
PLATE SUPPLY

or guidance signal varies. It is obvious that a high degree of reliability is required for controller valves in missile applications. Considerable development has gone into this high-demand component which might appear to the casual observer to be less important than many others in missile control and guidance systems.

Relays are also used as *controllers* where electrical circuitry must be switched to provide input to motors for actuator-operation. Although basically these relay applications are similar in principle to those for controlling the smallest servomotor for model aircraft, they are required to switch the high currents which are supplied to the motor-actuators described earlier. Very heavy contacts are provided and overload-fusing is employed in the event that heavy currents are drawn for more than a very short period of time.

Controller Amplifiers

Between the controller and the output of a missile guidance computer we find amplifiers in practically all control and guidance systems. Since the signals may be either a-c or d-c it is necessary to have certain auxiliary devices associated with these units. For instance some system components may require d-c and some may require a-c for their operation. Thus, it is necessary to convert signals from d-c to a-c and vice versa. Typical d-c to a-c converters are mechanical choppers, vacuum tube choppers and rectifier modulators.

A mechanical chopper operates as a synchronous switch. The d-c signal is simply broken at a desired rate by contacts attached to a vibrating reed. The vibrator-type power supplies which have been so widely used in automobile radio receivers (particularly, before the advent of transistors when higher plate voltages were required for tube amplifiers) provide a good example of a full-wave chopper in which contacts are made periodically to cause current flow in two directions through split primary windings of a power transformer. In a simple version of a vacuum-tube chopper the d-c signal is applied to the control grid, and the a-c chopping signal may be applied to the screen grid so that an alternating voltage related to the d-c input appears in the plate circuit. In the case of a rectifier modulator the a-c reference voltage controls a combination of diodes in an a-c bridge circuit in such a manner that current flows intermittently first in one direction and then in the other.

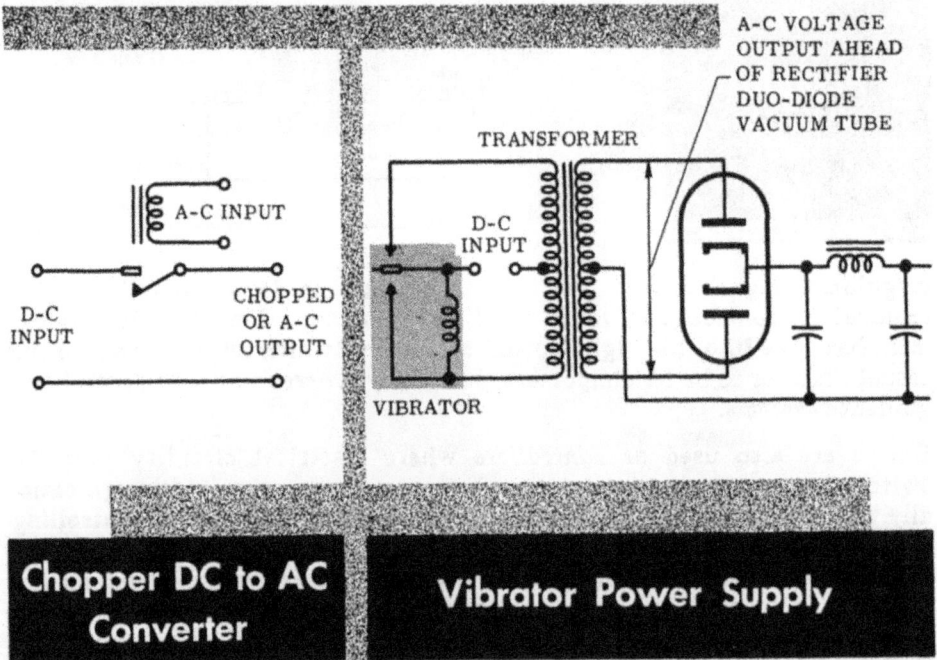

Chopper DC to AC Converter

Vibrator Power Supply

Controller Amplifiers (contd.)

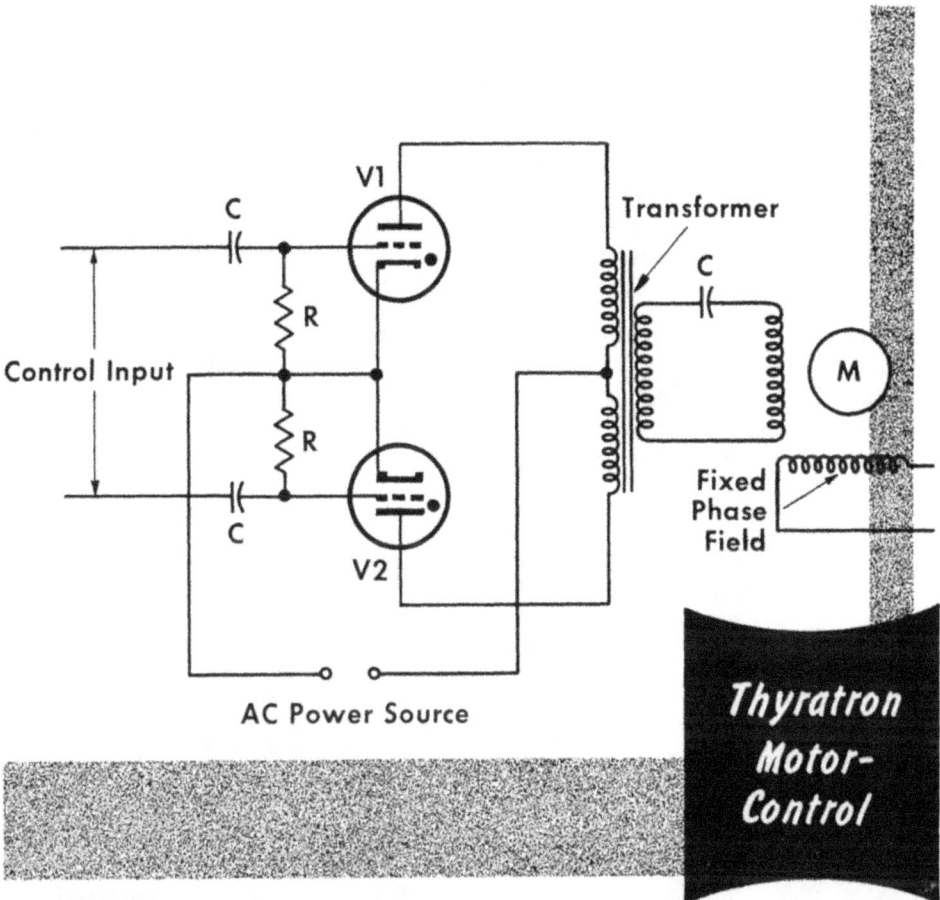

Conversion from a-c to d-c is basically that of normal power-supply techniques where rectification takes place, and a d-c voltage is derived from an a-c source. Of course such an effect can also be obtained by the use of commutating rotary machinery.

Both tube type and magnetic amplifiers are of interest in the servo-chain of missile control and guidance. Where large currents for motor control are to be handled by tube control, thyratrons can be applied. It is necessary to use gas tubes to handle the high current levels not practical for vacuum tubes.

A split-phase motor may be controlled through a thyratron circuit arrangement as shown above by supplying one field from a fixed source and the other through V1 and V2 in a push-pull arrangement. The level of the input voltage determines the amount and phase of the current which flows from the a-c source.

Magnetic Amplifiers

Magnetic amplifiers lend themselves to the control of large amounts of current because all of the power is carried through windings rather than inside the envelope of a vacuum tube. Well-known basic saturable reactor techniques are employed. Two coils L1 and L2 are wound on a common saturable core. L1 is connected in series with a variable d-c voltage so that sufficient current flows through it to place the operating point near a saturable level. In this condition the impedance of L2 can be varied considerably by small changes in the d-c supply level. Hence, an amplifying effect is realized in the impedance variation of L2 and the consequent variation of the current flowing through the output load.

The simple circuit shown below only illustrates the principle of the magnetic amplifier. In practice a rectifier is used in the a-c circuit to prevent the current from reversing and cancelling the effects of the control winding. Thus a so-called *half-wave amplifier* is realized. It is obvious that a d-c component would be present in the secondary in this arrangement.

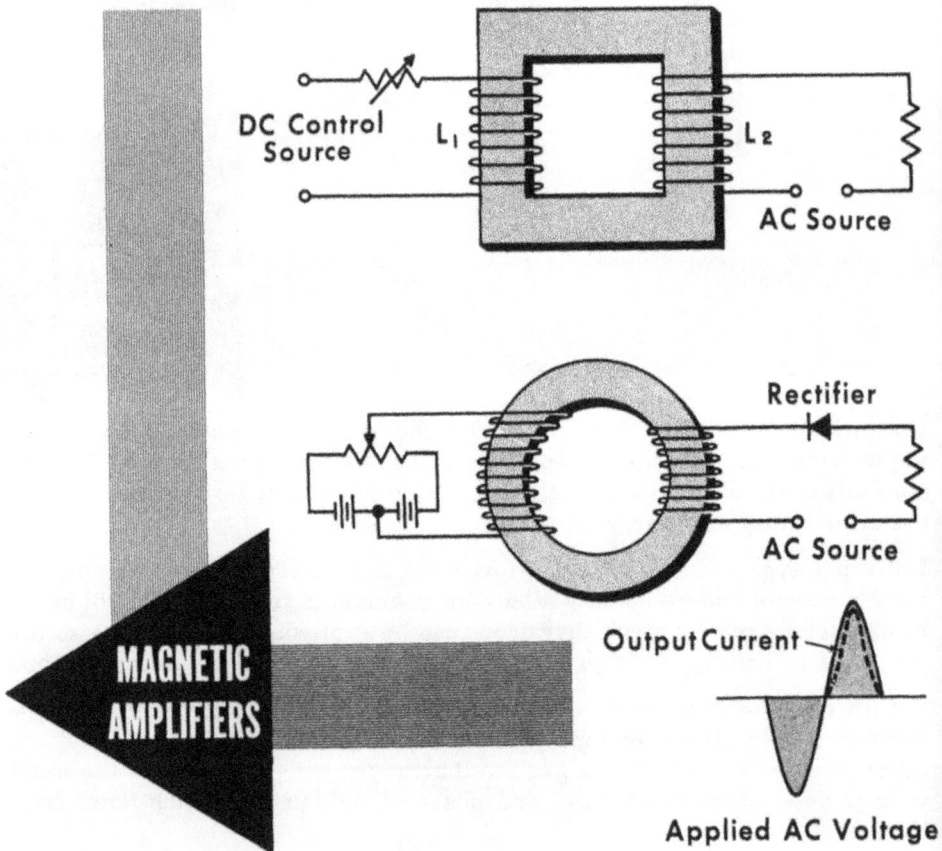

Magnetic Amplifiers (contd.)

Most magnetic amplifiers employ a full-wave circuit consisting of two saturable reactors A and B. When the a-c is of the proper polarity to cause rectification in the secondary winding of A, current flows in that circuit. On the

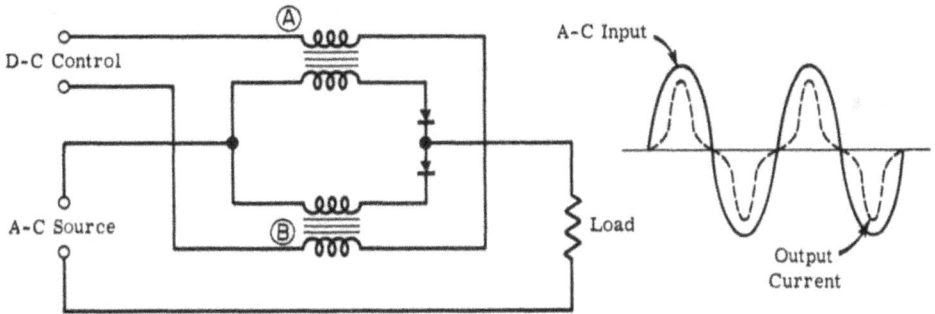

next alternation current flows through the secondary of B. Thus, both al-
ternations are carried through to the load and both are regulated by the
saturation produced in the d-c primary winding.

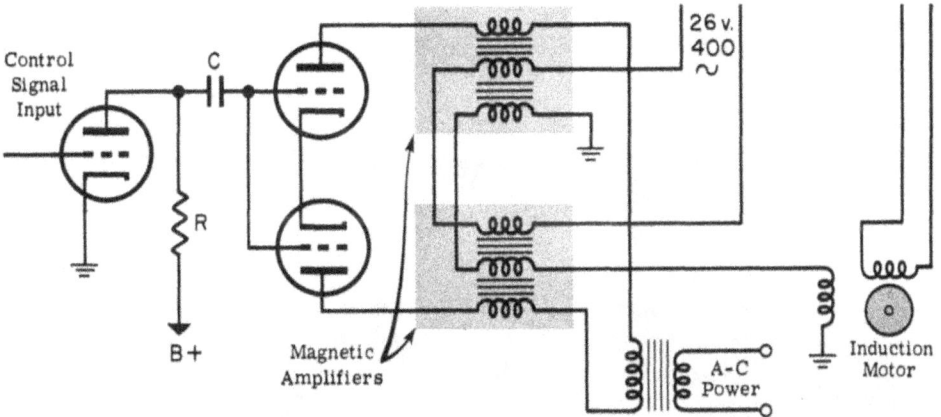

A typical magnetic amplifier for an autopilot is shown here. The a-c power
to be applied to one of the windings of an induction-actuator motor is fed
through primaries and secondaries, whose impedances are controlled by the
d-c flowing into two output tubes. Thus a combination of both magnetic
amplifiers and vacuum-tube amplifiers is employed.

Magnetic Amplifiers (contd.)

Magnetic amplifiers suffer primarily from time-lag limitations due to inductive effects and the fact that at least one-half cycle of the a-c excitation frequency must be effective for the load current to change to a new amplitude level. Thus, if a 400 cycle a-c excitation frequency is employed, the amplifier will be able to handle frequencies only at some point below 400 cycles. For instance, it may be able to handle up to only 250 cycles. There are also considerations of waveform distortion due to nonlinear characteristics and the matter of compactness and weight. Transistors have considerable advantages in regard to weight and compactness, and are being used more widely as their power handling and temperature capabilities are improved. They may be combined with magnetic amplifiers to obtain faster response through negative inductance compensation.

Another facet of a complete control or guidance channel is the follow-up amplifier. This unit feeds back to the servo amplifier a signal proportional to the actual output at the actuator. By so doing it tends to stabilize the system, just as a negative feedback channel in an audio amplifier tends to dampen oscillation or other violent swings of motion. A loop through the servo amplifier, transfer valve or controller, the actuator and the follow-up unit completes such a picture. A typical loop with a hydraulic system is shown.

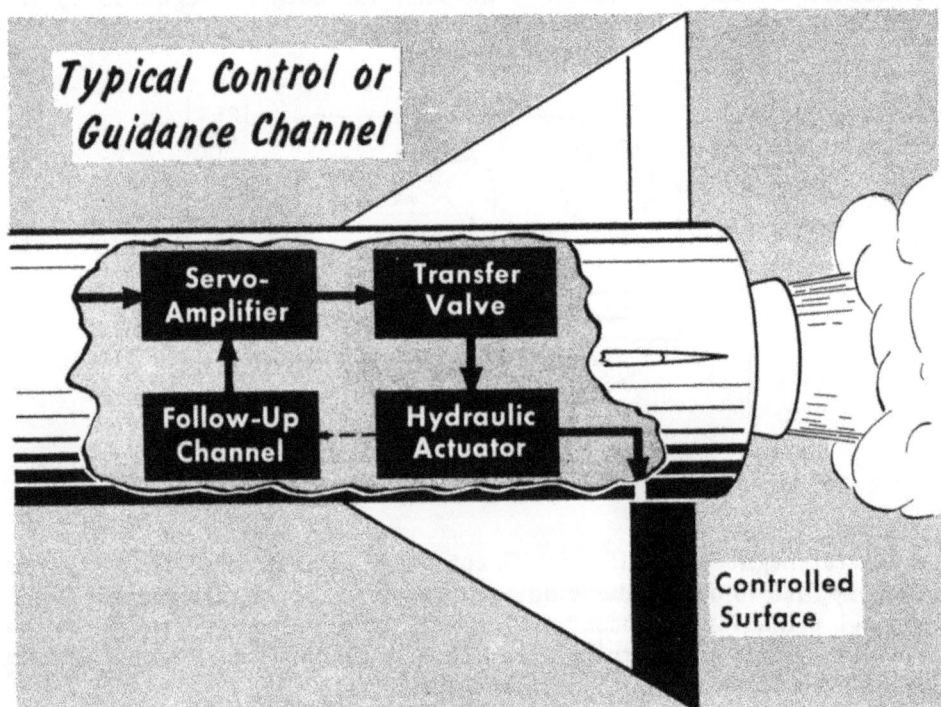

Typical Control or Guidance Channel

Auxiliary Power Supplies

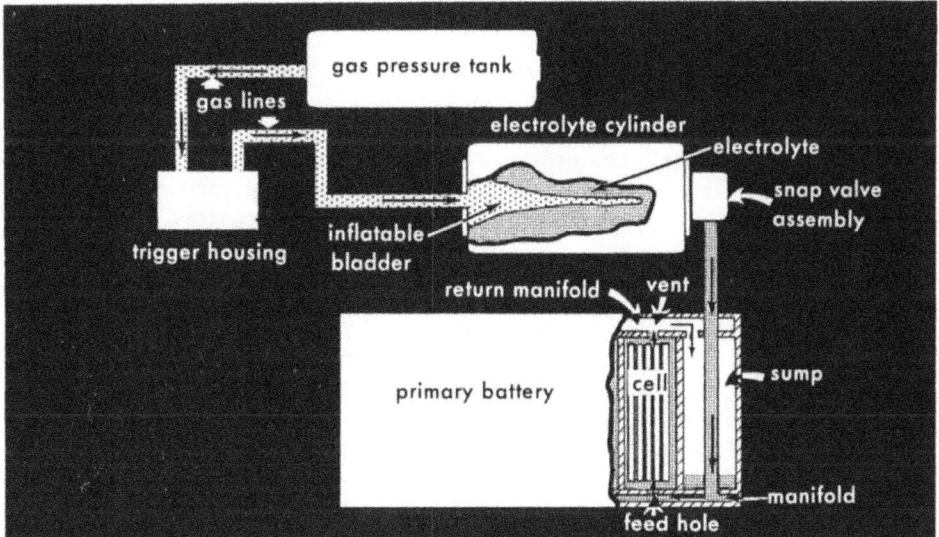

BLOCK DIAGRAM OF AUTOMATIC ACTIVATION SYSTEM

An important aspect of any control channel as well as the entire control and guidance system is the auxiliary power supply. In the case of the hydraulic-electric and pneumatic-electric systems, power sources for pressurizing the liquid or gas must be available as well as electrical power for all of the electronic portions. Such sources are called APUs — an abbreviation for auxiliary power supplies. Three basic types have been employed in a variety of missile designs — (1) batteries, (2) cold gas, and (3) hot gas systems. Requirements vary with the type of missile from short duration one-shot applications to longer duration higher level demands.

For smaller missiles in particular and some larger missiles, batteries have often been employed as the *APU*. They supply low voltage electrical power directly and can drive alternators and the hydraulic pump through a motor. Primary batteries, such as the silver-zinc type, are usually employed. Battery APUs generally have a long storage life because they are kept in an inert condition until missile launching time. Then the electrolyte is forced into the cells to activate them by discharging a carbon-dioxide or hot gas squib associated with it.

Although battery systems have been able to realize competitive power per pound and unit volume through design advances, much of these gains are lost due to the necessity of operating an electric motor for the hydraulic pump. Due to the weight of the motor it is not likely even with higher voltages that the battery APU will be able to compete with the hot gas APU and its lighter turbine.

ACTUATORS AND RELATED COMPONENTS

Auxiliary Power Supplies (contd.)

Cold gas APUs are usually employed for short-time applications to power only the hydraulic system. The prime power is supplied from a high pressure gas source. As the missile is launched, the power source is activated by an arming mechanism which punctures the sealed gas. In some of these systems the sealed-gas pressure is too high to use directly so it is followed by a regulator to reduce the pressure to a somewhat lower level. In its operation gas flows at the proper pressure to an accumulator-reservoir unit, where it powers the hydraulic fluid to the valves and actuators.

In hot gas APUs either monopropellant or bipropellant fuels are converted into high pressure gas in a combustion chamber. The gas is directed in a jet to a turbine wheel which drives both the hydraulic pressure pump and the alternator to supply electrical power to the electronic portion of the missile system. The system may be regulated by a hydraulic pressure feedback which acts to control the opening of the turbine nozzle. Electrical feedback may also be applied from the alternator through a magnetic amplifier to control the flow of fuel to the combustion chamber. Hot gas APUs offer the advantage of high energy per pound of fuel.

ACCESSORY POWER SYSTEM

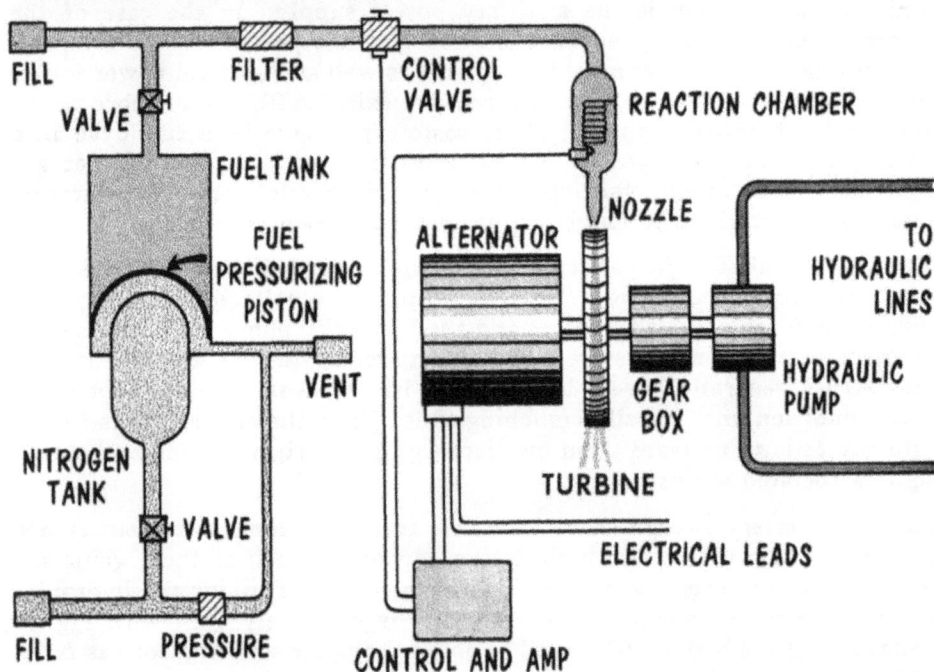

Hot Gas Servos

A new approach applicable in some cases to provide size and weight reduction is the use of hot gases directly in a servo control system. Instead of using hot gas pressure to power hydraulic fluid, with this method the hot gas is controlled directly by a valve to operate the actuator. Solid or liquid propellant fuels are burned in a gas generator to develop high-temperature gas at high pressure and a constant mass-flow rate.

The gas is fed directly to the actuator, and the flow is regulated by a so-called *flapper-nozzle* valve. Such a valve controls the flow of gas through nozzles attached to the system above the actuator pistons. By decreasing the gas flow through one nozzle and increasing the flow through the other nozzle, the pressure is increased on one of the actuator pistons, and reduced on the other. Such a pressure variation causes one piston to move down and the other to move up, thereby controlling the surface or vane element on the missile.

If the controlled element operates in the air, a direct load is placed on its surface and feedback is obtained automatically through the pressure system. For control elements not having the airload the need for an external feedback channel still exists. Although hot gas servos offer significant volume, cost, and weight reductions for certain applications, their effective use is dependent upon the required servo bandwidth, the stall torque-to-inertia ratio and the maximum angular deflection required by the controlled element.

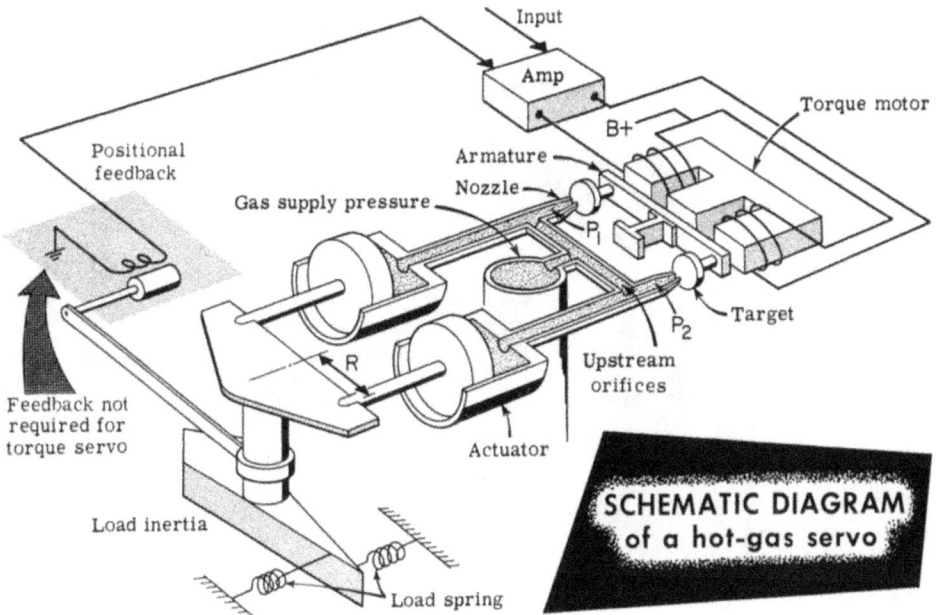

SCHEMATIC DIAGRAM of a hot-gas servo

Digital Servos

The availability of binary bits of information from digital computers has emphasized the need for servo units which can convert digital data into mechanical motion. Since this form of output is presented in a step-by-step sequence, the servo unit most suited to handle it will respond in a similar fashion. Stepper or digital motors offer a means for this direct conversion of electrical impulses into mechanical force for positioning flight and thrust elements, which can be motor-controlled.

In one design a basic motor employs two rotary solenoids to produce incremental motion of its output shaft in either direction. By energizing either of these solenoids a combination linear and rotational motion moves a ratchet axially into engagement with a mating ratchet. By so doing it imparts a given amount of rotation into an output shaft. Some designs of digital motors can handle up to 30 or 40 pulses per second, and maintain a given position without power consumption.

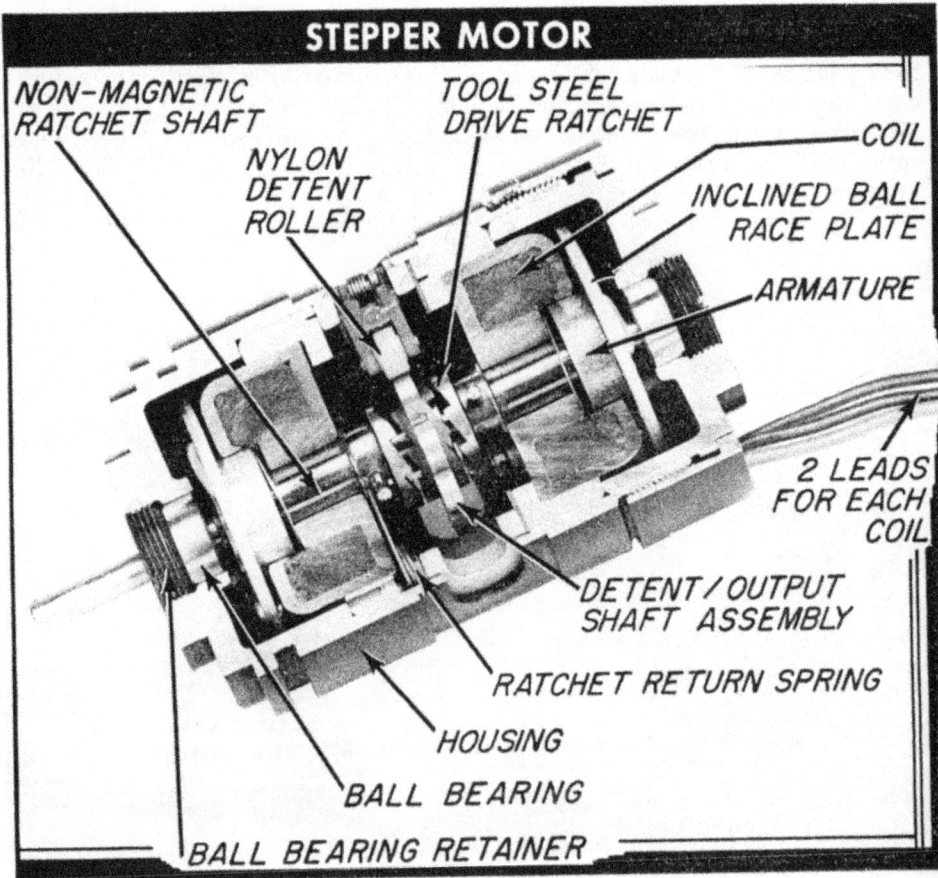

STEPPER MOTOR

NON-MAGNETIC RATCHET SHAFT — TOOL STEEL DRIVE RATCHET — COIL — NYLON DETENT ROLLER — INCLINED BALL RACE PLATE — ARMATURE — 2 LEADS FOR EACH COIL — DETENT/OUTPUT SHAFT ASSEMBLY — RATCHET RETURN SPRING — HOUSING — BALL BEARING — BALL BEARING RETAINER

GUIDANCE SYSTEMS IN TYPICAL PRESENT-DAY MISSILES AND DRONES

Missile Name and Class	Type of Guidance	Hydraulics Used For:		Pneumatics Used For:	
		Flight Control	Engine Control	Flight Control	Engine Control
AIR-TO-AIR					
Falcon	Semi-active homing and infrared	x	—	—	—
Sidewinder	Infrared homing	—	—	x	—
Sparrow I	Beam rider	x	—	—	—
Sparrow III	Semi-active radar homing	x	—	—	—
AIR-TO-SURFACE					
Bullpup	Command	x	—	—	—
Corvus	Passive radar homing	x	—	—	—
Crossbow	Passive radar homing	x	—	x	—
Hound Dog	Inertial	x	—	—	—
Wagtail	Inertial	x	—	—	—
SURFACE-TO-AIR					
Bomarc	Command active radar homing	x	x	—	x
Hawk I	Semi-active radar homing	x	—	—	—
Nike-Ajax	Command	x	—	x	x
Nike-Hercules	Command	x	—	—	—
Nike-Zeus	Command radar homing	x	—	—	—
Talos	Beam-rider semi-active homing	x	x	x	x
Terrier	Beam rider	x	—	—	—
SURFACE-TO-SURFACE					
Atlas	Radio-inertial	x	x	—	x
Corporal	Command	—	—	x	x
Jupiter	Inertial	x	—	x	x
Polaris	Inertial	x	—	—	—
Redstone	Inertial	x	—	x	x
Regulus I	Command	x	—	x	—
Sergeant	Inertial	—	—	—	—
Snark	Stellar-inertial	x	—	—	—
Thor	Inertial	x	—	x	—
Titan	Radar command inertial	x	—	—	—
DRONES					
Q-2	Radio-radar	—	—	—	—
Q-4A	Radar	x	—	x	—
Q-5	Command	—	—	—	—
RP-76	Radio	—	—	—	—

GLOSSARY

Accelerometer — A device that measures one or more components of the accelerations of a missile.

Actuator — A device for moving the control surfaces or thrust directors of a missile.

Aileron — A hinged or movable surface on an airframe, the primary function of which is to induce a rolling moment on the airframe. It usually is part of the trailing edge of a wing.

Amplifier, Controller — A device for increasing the signal level applied to the controller valve or relay.

Amplifier, Magnetic — A device employing saturable magnetic components as a means for achieving increased signal level.

Analog-to-Digital Converter — A device for converting measured data into binary bits of information suitable for digital data handling equipment.

Antimissile Missile — An explosive missile launched to intercept and destroy another missile in flight.

Attitude, Flight — The inflight position of a missile as described by giving the inclinations of its axes to the Earth or any other specified frame of reference.

Autopilot — An automatic control mechanism for keeping an aircraft or missile in level flight on a set course, or for executing desired maneuvers.

Ballistic Missile — A missile propelled and guided during the first portion of its flight but later traveling a natural and uncontrolled trajectory.

Beacon, Radar — Usually, a nondirectional radiating device, containing an automatic radar receiver and transmitter system, that receives pulses (interrogation) from a radar and returns a pulse or set of pulses (response).

Beam Rider — A guided missile, which follows a directed radio or radar beam pattern toward a target. Equipment is incorporated within the missile to determine its position relative to the center of the beam pattern.

Binary Arithmetic — A counting method employed in digital computers. Counting is accomplished in terms of the presence or absence of a pulse.

Celestial Guidance — The guidance of a missile or vehicle by reference to celestial bodies.

Circular Error Probability — The radius of the circle within which half of the missiles aimed at the center may be expected to fall. Abbreviated as CEP and used primarily in regard to ballistic missiles.

Control, Closed-Loop — A control circuit in which a signal is fed back from the controlled element to the input point.

Control, Open-Loop — A control circuit in which there is no feedback from the controlled element to the input of the circuit.

Control, Proportional — Control in which the action to correct an error is made proportional to that error.

Control System — A system in a ballistic or guided missile that maintains attitude stability during powered flight.

Controller — An element for regulating the flow of power to an actuator in accordance with a controlling signal.

Comand Guidance — A guidance system wherein intelligence transmitted to the missile from an outside source causes the missile to traverse a directed path in space.

Computer, Analog — A computing device that works on the principle of measuring, as distinguished from counting.

Computer, Digital — A computer that functions on the principle of counting rather than measuring.

GLOSSARY

Computer, Trajectory Error — A computer which compares the actual trajectory of a missile with the one precalculated to intersect a desired target.

Coriolis Force — A deflecting force exerted by the rotation of the Earth upon any object in motion about it.

Differentiator — A circuit or device which produces an output substantially in proportion to the rate of change of an input quantity.

Doppler Effect — The apparent change in the frequency of vibrations, such as sound, light, or radar, when the observed and the observer are in relative motion to one another.

Elevator — A movable auxiliary airfoil, used to impress a pitching moment on the aircraft or missile.

Engine, Gimbaled — A rocket engine mounted on a gimbal: that is, on a contrivance having two mutually perpendicular and intersecting axes of rotation, so as to obtain pitching and yawing corrective motions.

Engine, Vernier — A rocket engine of small thrust used to obtain a fine adjustment in the velocity and trajectory of a ballistic missile.

Error Signal — In tracking systems a voltage, depending upon the signal received from the target, whose sign and magnitude depend on the angle between the target and the center of the scanning beam.

Frequency Division — The division of a given frequency band into a number of channels for conveying a corresponding number of separate and distinct signals.

Gating — An opening and closing of vacuum tube, transistor, or other devices for passing or blocking signals or pulses.

Gimbal — A mechanical frame containing two perpendicular intersecting axes of rotation. In the case of a gyroscope the rotor is housed inside one or more gimbals. The rotor-gimbal assembly precesses in accordance with applied external torques and signals are provided for correcting for these torques.

Guidance System — A system associated with a guided missile for establishing a desired path from launch to target.

Guided Missile — A missile directed to its target while in flight or motion, either by a preset program, or a self-reacting device within the missile, or by radio or radar command outside the missile. Combinations of same may be employed.

Generator, Torque — An arrangement associated with a gyroscope gimbal to provide a torque to correct for precession.

Gyroscope — A disc or mass, arranged to spin rapidly about its axis, and also free to rotate about one or both axes perpendicular to each other and the the axis of spin. An external applied torque will cause it to change its spin axis in a direction perpendicular to that of the applied torque.

Homing Guidance — A system by which a missile steers itself toward a target by means of a self-contained mechanism, activated by some distinguishing characteristic of the target.

Hydraulic System — A system which controls actuators through the flow of pressurized fluids.

Inertial Guidance — A self-contained guidance system where accelerometers provide data to guide the missile without resort to outside information. They are associated with suitable computers and may be mounted on a stable platform.

Infrared — Energy having a wavelength longer than that of visible light, and associated with heat sources.

Integrator — A device or circuit whose output is proportional to the product of the instantaneous applied inputs and their duration.

Modulation, Pulse — The process of varying the amplitude, frequency, position, duration, or number of pulses with time to transmit information.

Platform, Stable — A platform on which gyroscopes are mounted to obtain signals for maintaining its position in a fixed relationship in space.

GLOSSARY

Power Supply, Auxiliary — A power supply associated with a missile to provide electrical or hydraulic power (abbreviated APU).

Precession — A change of the axis of a rotating body, such as a spinning projectile or gyroscope, the effect of which is to rotate this axis (axis of spin) about a line (axis of precession) perpendicular to the original direction and to the axis (axis of torque) of the moment producing that change.

Proximity Fuze — A device designed to initiate a detonation when its carrier passes with a certain range of proximity to a target.

Radio or Radar-Inertial Guidance — A command type of guidance system. It includes a radar tracking unit on the ground, a radar beacon in the missile, a computer that receives missile position and velocity information from the tracking unit and furnishes appropriate signals to the command link to steer the missile, a command link between the ground and the missile usually incorporated in the tracking radar system, and an inertial system for partial guidance particularly in the absence of radar signals.

Rudder — A hinged or movable auxiliary airfoil on an aircraft or missile, the function of which is to impress a yawing moment on the aircraft.

Schuler Tuning — A method for simulating a pendulum having a hypothetical length equal to the radius of the earth and thereby determining the local vertical in an inertial navigator moving around the Earth.

Servo-System — A closed-loop automatic-control system so designed that the output element or output quantity follows, as closely as desired, the input to the system. The output is caused to follow the input by the action of the servocontroller upon the output element in such a way as to cause the instantaneous error, or difference, between the output and input to approach zero. All servo systems are dynamic systems containing at least one feedback loop which provides an input signal proportional to the deviation of the actual output from the desired output. This property distinguishes servo systems from open-loop automatic-control systems.

Stellar-Inertial Guidance — An inertial guidance system supplemented with celestial tracking data. Also known as astro-inertial.

Time-Division — The spacing of separate pulse trains along a time axis to convey separate information signals. Such pulse trains may be interspersed with each other.

Tracking Radar — A radar system which locks onto the reflected signal from a target or missile, or the radar beacon signal of a missile and tracks the object reflecting or radiating the signal.

INDEX

INDEX

www.ingramcontent.com/pod-product-compliance
Lightning Source LLC
Chambersburg PA
CBHW020207200326
41521CB00005BA/277